Concise
Tree
Guide

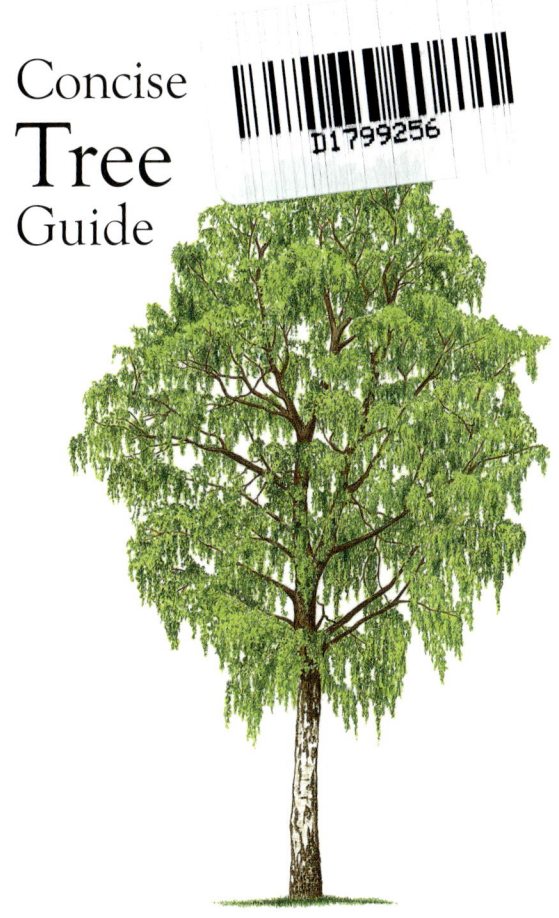

There are 47 individual Wildlife Trusts covering the whole of the UK and the Isle of Man and Alderney. Together The Wildlife Trusts are the largest UK voluntary organization dedicated to protecting wildlife and wild places everywhere – at land and sea. They are supported by 791,000 members, 150,000 of whom belong to their junior branch, Wildlife Watch. Every year The Wildlife Trusts work with thousands of schools, and their nature reserves and visitor centres receive millions of visitors.

The Wildlife Trusts work in partnership with hundreds of landowners and businesses across the UK in towns, cities and the wider countryside. Building on their existing network of 2,250 nature reserves, The Wildlife Trusts' recovery plan for the UK's wildlife and fragmented habitats, known as A Living Landscape, is being achieved through restoring, recreating and reconnecting large areas of wildlife habitat. As well as protecting wildlife this is helping to safeguard the ecosystems that we depend on for services like clean air and water.

The Wildlife Trusts are also working to protect the UK's marine environment. They are involved with many marine conservation projects around the UK, often surveying and collecting vital data on the state of our seas. Every year they run National Marine Week in August – a two-week celebration of our seas with hundreds of events taking place around the UK.

All 47 Wildlife Trusts are members of the Royal Society of Wildlife Trusts (Registered charity number 207238). To find your local Wildlife Trust visit wildlifetrusts.org

Concise

Tree

Guide

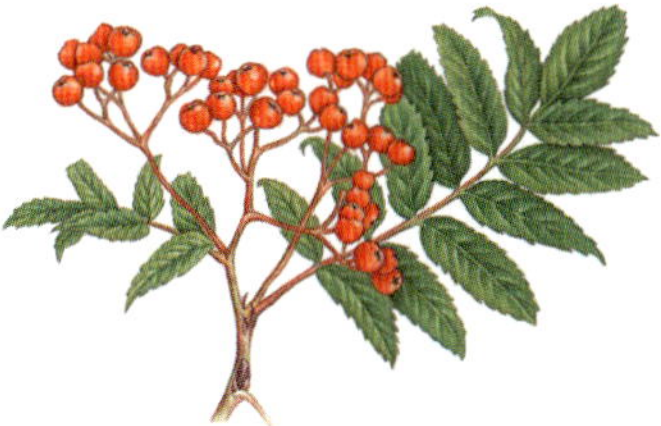

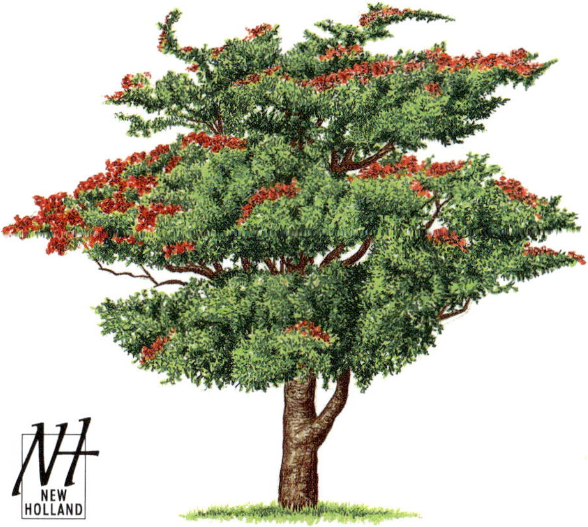

NH
NEW
HOLLAND

First published in 2011 by New Holland Publishers (UK) Ltd
London · Cape Town · Sydney · Auckland
www.newhollandpublishers.com
Garfield House, 86–88 Edgware Road, London W2 2EA, UK
80 McKenzie Street, Cape Town 8001, South Africa
Unit 1, 66 Gibbes Street, Chatswood, New South Wales 2067, Australia
218 Lake Road, Northcote, Auckland, New Zealand

10 9 8 7 6 5 4 3 2 1

ISBN 978 1 84773 605 5

Series Editor: Krystyna Mayer
Design: Alan Marshall
Illustrators: Cy Baker, Joyce Bee, Stuart Carter, Dan Cole, David Daly,
  Sandra Doyle, Bridgette James, Bob Press and Lyn Wells
Production: Melanie Dowland
Publisher: Simon Papps
Editorial Direction: Rosemary Wilkinson

The publishers thank Shirley Hawkins of The Wildlife Trusts for reading the text.

Reproduction by Modern Age Cp. Ltd., Hong Kong
Printed and bound in China by Leo Paper Group

Contents

Introduction

Trees are a major feature of a landscape. As well as being things of beauty, they bring a sense of scale and permanence to their surroundings, and mark the cycle of the seasons. They play a vital role in maintaining and protecting our environment: they prevent erosion, are part of the living system that recycles water, oxygen and carbon dioxide into the atmosphere, and provide habitats and food for many other organisms. Some also have economic value as timber or fruit trees, and as decorative ornamentals.

About 250 tree species are native to Europe, many of which are rare or have a restricted distribution. Over the centuries many more trees have been introduced for ornamental purposes, especially in cities and parks, and several species of conifer have been introduced for commercial purposes. Some species have escaped into the wild to spread and become naturalized. Many of the most familiar trees that can be seen growing in a country are therefore not indigenous to it. Only 35 species are truly native to the British Isles, for example.

The *Concise Tree Guide* describes the key identifying characteristics of more than 150 of the most interesting and widely encountered tree species, whether native or introduced, wild or planted in streets, parks or forestry developments. Native trees tend to support a rich array of wildlife that uses them as habitats – for sheer numbers of creatures harboured, the Pendunculate Oak (*Quercus robur*) cannot be beaten in Europe. Examples of the associations of some trees with other species, such as mammals, birds, insects and fungi, are provided for many of the trees featured in this book. The place of origin for each tree is also given.

Exploitation of Trees

The extensive primeval forest that once covered much of Europe now survives in only a very few places, such as the Bialowieza Forest in Poland. Most forests have been cleared and replanted.

Humans began clearances of forests about 6,000 years ago to make way for agriculture, creating a huge impact on the environment. In modern times, reafforestation programmes have led to large areas of land being returned to forest, although the trees planted are often not native but foreign species that yield timber.

Once our ancestors realized that trees produced a material that had uses other than firewood, they were ingenious in their production of useful crops. By coppicing and pollarding they both increased the lives of trees, and produced wood of lengths and shapes that were suitable for a variety of uses. Coppicing of hazel every 5–6 years, for example, produced rods of suitable width to be used as laths in the wattle and daub walls of houses.

In south-west Spain the *dehesa*, a wood-pasture, has been managed in the same way for at least 2,000 years. Here an extensive farming system depends on the delicate balance between trees, livestock and a rotational farming regime. The Holm Oaks (*Quercus ilex*) that grow in the *dehesa* are pruned to have three main branches with no crown, so that maximum sunlight reaches the forest floor, encouraging the growth of flowers. The flowers attract the bees that pollinate the trees, thus providing acorns on which pigs feed. This is a sustainable farming system, albeit one that produces a standard of living for the farmer scarcely above subsistence.

Tree Management & Conservation Today
Woodland management declined in many places in the 20th century, and its revival over the last 25 years has in a great part been due to nature conservationists. They realized that traditional woodland management created the mixture of sizes and ages of trees that provided the conditions needed for the survival of mammals, birds, insects, flora and fungi. The Wildlife Trusts across Britain have many woodland nature reserves and are leaders in the management of woodlands to produce the greatest quantity of wildlife.

What is a Tree?
The definition of a tree is a woody perennial plant with a main stem or trunk generally 6m or more in height, and usually a distinct crown. Since trees take a great number of years to reach maturity, many never reach their maximum height because of human interference such as coppicing and pollarding.

Like all organisms, plants have been divided into taxonomic groups, the most important of which are the class, the order, the family and the species. In addition to the common English name there is the scientific name of the species. The scientific name contains two names: the first is the genus and is shared by several species, while the second is the specific name. Thus, the Scots Pine is *Pinus sylvestris*, while the Black Pine is *Pinus nigra*. However, there are two varieties of Black Pine, one from Corsica and one from the Pyrennees. They are described as *Pinus nigra* var. *pyrenaica* and *Pinus nigra* var. *maritima*.

Geographically isolated populations of species may develop their own characteristics and eventually become separate species that are unable to breed with other geographically separate populations of the original species. Human interference has confused this due to the interbreeding of different varieties, and interbreeding can lead to difficulties with identification. Hybrids may be between species, or between varieties of the same species. A hybrid tree may look sufficiently like either of its parents as to be indistinguishable.

The third type of hybridism is inter-generic, where trees of different genera interbreed. The less closely the two species are related the more vigorous the resulting hybrid will be, growing higher and probably bearing more fruits. An excellent example of a tree bred as a hybrid is the Leyland Cypress, which is considered by some to be a pest. It was bred from two North American species, the Monterey Cypress (*Cupressus macrocarpa*) and the Nootka Cypress (*Chamaecyparis nootkatensis*). As a hybrid, the Leyland Cypress's scientific name is prefixed by a cross – thus, x *Cupressocyparis leylandii*.

Unlike in the animal kingdom, hybrid plants can produce fertile offspring, although the hybrid vigour diminishes within a few generations.

Identifying Trees

Trees are sometimes difficult to identify because, unlike in other plants, tree parts are often out of reach. Identification is a deductive process based on single features or a combination of features. From the shape it may be possible to identify the genus. The leaves, flowers and bark may all be diagnostic. In woods it may not be possible to see the shape clearly, and the leaves and bark become more important in recognizing a tree. In winter deciduous trees will not be in leaf, but the size, shape and structure of fallen leaves will still be an essential aid to identification.

Bark

Forming the protective outer layer of the trunk and branches, bark can be thin and smooth, or thick, rough and corky. It splits as the tree grows, cracking or peeling away to be constantly replaced by new bark. The pattern of cracks and colour of bark or of areas revealed as it falls can be useful recognition features.

Twigs & Branches

Twigs and branches make up the crown of a tree. Its overall shape can vary between trees of different species, and between those of the same species but of different ages. Young cedars, for example, are conical, but old ones are flat-topped and spreading.

Leaves

Most trees have leaves alternating on the shoot, but some have opposite pairs or whorls of leaves. The leaves may be divided into leaflets. **Pinnate leaves** have the leaflets in two parallel rows with sometimes a leaflet at the tip. **Palmate leaves** have the leaflets radiating from the leaf stalk like the fingers of a hand. Similarly, the leaves may be pinnately or palmately lobed without being completely divided into leaflets. The leaf margin may be entire or variously toothed. Leaves also vary in colour, texture and hairiness. They are usually paler and more hairy on the underside. Conifers have leathery leaves that are either narrow and needle-like, or scale-like, overlapping and pressed against the shoot. Most, but not all, are evergreen.

   Deciduous trees shed all their leaves annually. In Europe this occurs in autumn, when the dying leaves may present a striking range of colours, mainly shades of yellow, orange and red. New leaves are produced each spring. **Evergreens** shed their leaves, but more gradually than deciduous species, constantly replacing them so that there is always a full crown of foliage.

Flowers

Trees bear flowers, although in temperate Europe, where insects are a major means of pollination, they are smaller and less obvious than flowers in the tropics and sub-tropics, where birds and tree-climbing mammals have a role in pollination. A single flower may be male or female, or both. It may be simple or complex, and appears in many forms, such as the catkins of hazels and pea-blossom flowers of acacias. In **dioecious species** all the flowers of an individual tree are of one sex; in **monoecious species** such as conifers, there are male and female flowers on each tree.

Flowers consist of sepals, petals, stamens and ovaries. The number, size and colour of the petals and sepals provide most clues to identity. Stamens are the male parts of a flower, ovaries the female parts. Flowers may be male, female or both (hermaphrodite).

Cones

Conifers have male and female cones instead of flowers and fruits. Male cones are yellow when shedding pollen, but it is the large female cones that are most noticeable. When ripe they are generally woody, but can be fleshy and berry-like, as in Junipers.

Fruits

There are two broad categories of fruits in non-coniferous trees, fleshy and juicy, and dry. Fleshy and juicy fruits include all berries and berry-like fruits, as well as firm fruits such as apples. Dry fruits include pods, capsules and nuts.

Tree Types

The species in this book are grouped into families, the essential characteristics of which are described below.

Taxaceae/Yews Evergreen needles and fleshy, berry-like fruits (red 'arils'). There are six similar species.

Ginkgoaceae/Gingkos Deciduous fan-shaped leaves. An ancient family with just one surviving species.

Cupressaceae/Cypresses Evergreen trees or shrubs. Most have overlapping and scale-like leaves, but some junipers have needles as well as or instead of scale-leaves. Cones are globular with scales touching edge to edge. In junipers they are fleshy.

Taxodiaceae/Redwoods Massive evergreen trees, three species of which grow in Britain. They can live for more more than 3,000 years, and originate in the Americas and China (Dawn Redwod).

Pinaceae/Pines Evergreen or deciduous trees with needles in bundles with a sheathing base. Cones are egg shaped.

Salicaceae/Willows & Poplars Deciduous trees or shrubs. Flowers are borne in male and female catkins on separate trees. Seeds are tufted with white hairs. All species like rich and wet soil.

Juglandaceae/Walnuts Deciduous trees that exude latex from broken twigs. Leaves are pinnately divided. Flowers are in separate male and female catkins on the same tree.

Betulaceae/Birches & Hazels Deciduous trees or shrubs with broad alternate leaves. Flowers are in separate male and female catkins. In birches the small winged nutlets are borne in cone-like catkins. Hazels have large woody nuts.

Fagaceae/Beeches & Oaks Deciduous or evergreen trees with alternate, often lobed or spiny leaves. Flowers are small, the males in catkins, the females in clusters of 1–3. Nuts or nut-like fruits are partly or wholly enclosed in a scaly or spiny cup.

Ulmaceae/Elms Deciduous trees. Small flowers are either all hermaphrodite or with some males. Seed is surrounded by a papery wing.

Moraceae/Mulberries Deciduous with toothed or deeply lobed leaves. Male and female flowers are sometimes borne on separate trees. Fruits are berry-like.

Magnoliaceae/Magnolias Evergreen trees or shrubs with alternate, entire leaves. Large and solitary flowers.

Lauraceae/Laurels Usually evergreen trees and shrubs. Mostly tropical, although a few hardy species occur in northern areas. Many species are aromatic.

Platanaceae/Planes Deciduous trees with often peeling bark and mostly deeply palmately lobed leaves. Leaves are alternate. Separate male and female 4–6-petalled flowers.

Rosaceae/Roses Deciduous or evergreen trees or shrubs with entire, lobed or pinnately divided alternate leaves. Usually showy flowers are five-petalled. Fruit is fleshy and firm or juicy, with one to several stony seeds.

Leguminosae/Peas Deciduous or evergreen trees, often with pinnately divided leaves. Fruit is a dry and slender pod with few to many seeds.

Aceraceae/Maples Deciduous trees with deeply palmately lobed leaves, which are paired. Usually five-petalled hermaphrodite flowers; winged fruits form pairs.

Hippocastanaceae/Horse Chestnuts Large deciduous trees. Leaves are palmately divided into leaflets. Large flowers have 4–5 petals and long, curved stamens. Fruit is one to several nuts in a spiny case.

Aquifoliaceae/Hollies Evergreen trees with spiny leaves. Flowers are white and fragrant, four-petalled, and with males and females growing on separate trees. Fruit is a berry.

Celastraceae/Spindles Deciduous trees with paired leaves and twigs. Inconspicuous flowers. Fruit is a capsule.

Buxaceae/Boxes Evergreen trees with paired leaves and twigs. Inconspicuous, separate male and female flowers. Fruit is a capsule.

Rhamnaceae/Buckthorns Small deciduous trees or shrubs. Frequently spiny or thorny. Leaves are paired or alternate. Flowers are hermaphrodite, or males and females on separate trees, and have 4–5 petals. Fruit is berry-like or a hard nut.

Tiliaceae/Limes Deciduous trees with alternate heart-shaped leaves. Fragrant five-petalled flowers hang in clusters beneath a wing-like bract. Fruit is a small nut.

Tamaricaceae/Tamarisks Small and slender trees or shrubs with wand-like twigs and tiny and scale-like, clasping leaves. Small flowers borne in dense heads. Seeds are dispersed by wind.

Elaeagnaceae/Oleasters Small deciduous trees and shrubs that are often spiny. Leaves are paired or alternate. Fragrant flowers have 2–4 sepals. Fruit is berry-like or a hard nut.

Myrtaceae/Myrtles Evergreen trees with peeling, shredding or fibrous bark and two kinds of very aromatic leaves: juveniles are paired and often blue; adults alternate and dull green. Flowers are enclosed in capsules with a cap that is eventually shed. Fruit is a woody capsule.

Lythraceae/Loosestrifes Small trees, shrubs and herbs with leaves that are usually in pairs, and red, pink or purple flowers. Most species are found in the tropics, but some also grow in temperate areas.

Cornaceae/Dogwoods Small deciduous trees or shrubs. Leaves are paired and entire. Small and clustered four-petalled flowers. Fruit is a berry.

Ericaceae/Strawberry Tree & Rhododendron Evergreen or deciduous trees or shrubs with alternate leaves. Flowers are variable, but the majority have five petals that are joined at the base.

Oleaceae/Olive & Ashes Evergreen or deciduous trees or shrubs. Leaves are paired and entire, or pinnately divided into toothed leaflets. Four petals, if present, are joined in a tube. Fruit is either a berry or dry and winged.

Bignoniaceae/Bignonias Deciduous trees with large paired leaves. Flowers are tubular and have five petals. Pod-like fruit is long.

Caprifoliaceae/Elders Deciduous or evergreen small trees or large shrubs. Paired leaves are pinnately divided, deeply lobed or undivided. Five-petalled flowers are borne in large clusters. Fruit is a berry.

Agavaceae/Century Plants Evergreen trees with a single straight or forked trunk, but no branches. Leaves are large and sword-shaped. Flowers emerge in midsummer on long and tall spikes. Fruit is a small rounded berry that contains several black seeds.

Arecaceae/Palms Evergreen trees with a single straight or forked trunk, and no branches. Leaves are fan- or feather-shaped, and are characteristically all crowded at the top of the trunk.

Trees, perhaps more than any other organism, have been moved around the world by humans and have been interbred. This may make identification difficult, but it does give you a chance to see many exotic species without having to travel the four corners of the world.

Yew
Taxus baccata

DESCRIPTION Height 10–25m. Evergreen. One of the world's longest lived trees, with individuals more than 2,000 years old growing in some English, Welsh and northern French churchyards. Crown in young trees tends to be cone-shaped, becoming columnar, then domed. Bark purple-brown, becoming scaly. Leaves deep green with two paler bands on undersides, 2–3cm long. All parts of tree, except for red aril, are poisonous to humans and livestock.

FLOWERS AND FRUITS Male and female flowers grow on separate plants. Male flowers are yellow; female flowers greenish. Flowers appear on shoots of previous year in late spring. Male cones in leaf axils. Fruit a red aril 1cm long, with a black seed.

HABITAT Throughout Europe as far as North Africa. Shade tolerant, and common in woods and scrub, especially on limestone. Grown in parks and gardens. Commonly used to create topiary and hedging, and clippings are used for anti-cancer drugs. Traditionally used to make English longbows.

Maidenhair Tree
Gingko biloba

Description Height 15–25m. Deciduous. Long-lived. Crown columnar for many years, becoming spreading and open. Leaves fan-shaped, two-lobed, 5–7 x 5–7cm.

Flowers and fruits Male and female flowers grow on separate trees. Male has small green-yellow catkins, 6–8cm long; female flowers small, stalked, round and knob-like. Fruit resembles a green plum, 2.5–4cm long, with a white edible seed.

Habitat Native to eastern China, and grown widely for its seeds in eastern Asia. Widespread in gardens as an ornamental and in towns as an amenity tree.

Deodar
Cedrus deodara

DESCRIPTION Height to 40m. Evergreen and coniferous. Distinguished from other cedars by its conical shape, drooping branches and longer leaves. Bark dark grey-brown. Needles soft, 3.5–4.5cm long.

FLOWERS AND FRUITS Male flowers erect, releasing yellow pollen in autumn; female flowers green. Ripe cones, 8–13cm long, dark brown.

HABITAT Native to the Himalayas, favouring dry mountainous areas. Widely planted as an ornamental.

Atlas Cedar
Cedrus atlantica

DESCRIPTION Height 20–35m. Evergreen and coniferous. Crown broadly conical at first, becoming wider with age. Bark dark grey. Branches ascending. Leaves needle-shaped, blue-green or dark green, 1.5–2cm long, growing singly on current year's shoots.

FLOWERS AND FRUITS Male flowers pinkish-yellow; female flowers green. Male and female cones open in autumn.

HABITAT Native to the Atlas Mountains. Now widespread as an ornamental on a variety of free-draining soils.

Western Red Cedar
Thuja plicata

DESCRIPTION Height 25–45m. Evergreen and coniferous. Fast-growing tree. Bark cinnamon-red, turning greyish. Leading shoot almost erect. Leaves bright green on top, lower sides paler, very fragrant, 3mm long.

FLOWERS AND FRUITS Female and male flowers borne at ends of short branches. Male flowers reddish; female flowers yellowish-green. Leafy cones, about 1cm long, mature in one year.

HABITAT Native to North America. Grows in moist acid soil and swamps in the wild. Popular tree, also grown for forestry.

Cedar of Lebanon
Cedrus libani

DESCRIPTION Height 20–40m. Evergreen and coniferous. Broadly rounded with a characteristic shelf-like flat top in mature trees. Massive branches grow horizontally. Crowns of young trees conical. Bark smooth, brown or dark grey at first, darkening with age and becoming cracked. Leaves are dark green needles, 1–3cm long.
FLOWERS AND FRUITS Male flowers release yellow pollen in autumn; female flowers coloured pale green. Cones, 7–10cm long, erect and barrel shaped.
HABITAT Native to Lebanon, Syria and southern Anatolia, growing on mountain slopes, but now extremely rare in its natural habitat. In Britain widely planted as a pollution-resistant ornamental and park tree.

Monterey Cypress
Cupressus macrocarpa

Description Height to 40m. Evergreen and coniferous. Markedly upswept branches with ropey or cord-like foliage. Crown narrow and pointed when young, but broadly domed in old trees. Bark yellowish-brown, ridged. Leaves blunt and scaly, 1–2mm long.

Flowers and fruits Cones, 3cm long, like footballs with 8–14 scales, each with a pointed central boss.

Habitat Native to southern California. Salt-resistant; used for coastal shelter and ornamental planting in western and southern Europe.

Italian Cypress
Cupressus sempervirens

Description Height to 30m. Evergreen and coniferous. Low and spreading in wild form, but more commonly dense and spire-like in cultivated form. Bark greyish, often with spiral ridges. Leaves scaly and blunt, 0.5–1mm long.
Flowers and fruits Cones, 3cm long, ellipsoid-oblong with 8–14 scales, each with a short blunt central point and often wavy edges.
Habitat Native to the Aegean. Widely planted in southern Europe and naturalized in many places. In Britain hardy to north-east Scotland, and planted as an ornamental.

Lawson Cypress
Chamaecyparis lawsoniana

DESCRIPTION Height 20–40m. Evergreen and coniferous. Bark at first cinnamon-red, turning rich dark brown. Leading shoot is drooping and small leaves are 2mm long.

FLOWERS AND FRUITS Female flowers grow on ends of small branchlets; male flowers grow on ends of branches and have black scales. Cones, up to 8mm across, are spherical.

HABITAT Native to North America, growing on ridges and in valley slopes in the wild. Tolerates a variety of soils. Widespread ornamental.

Western Hemlock
Tsuga heterophylla

DESCRIPTION Height 40–50m. Evergreen and coniferous. Leading shoot arches widely so tip points downwards, giving the tree a drooping appearance. Young stems are hairy. Needles taper to blunt tips, and vary in length from less than 0.5cm to about 2cm.

FLOWERS AND FRUITS Flowers in late spring on previous year's growth. Male cones red. Mature cones 2–3cm long, light brown with a few rounded scales.

HABITAT Native to the Pacific seaboard of North America, growing in moist acid areas, on lower slopes. Will grow in shade. Grown commercially for paper pulp. Brought to Britain during the 19th century as a decorative specimen.

Phoenician Juniper
Juniperus phoenicea

DESCRIPTION Height to 8m. Evergreen and coniferous. Small tree or sometimes spreading shrub. Foliage of two distinct kinds: cord-like shoots with scaly leaves, and young growth bearing needles. Leaves of young growth up to 14mm long, needle-like and wide-spreading. Scaly adult leaves only 1mm long, blunt with pale margins.

FLOWERS AND FRUITS Cones, 6–14mm long, berry-like.

HABITAT Widespread throughout coastal Mediterranean region. Included in some northern European collections.

Juniper
Juniperus communis

DESCRIPTION Height to 6m. Evergreen and coniferous. Can also grow as a low-growing twisted shrub with spreading branches. Bark a rich reddish-brown. Leaves are short spreading needles, 2cm long, spiky blue-green, borne in whorls of three.

FLOWERS AND FRUITS Male and female flowers grow on separate plants. Male flowers yellow; female flowers green. Berries green in first year, ripening to dark purple in second.

HABITAT Scattered throughout Europe, growing in a variety of soils, although it prefers lime-rich soil.

Wellingtonia
Sequoiadendron giganteum

DESCRIPTION Height 50m. Evergreen and coniferous. Noted for its height, bulk and longevity. Can live for up to 1,500 years; up to 3,500 years disputed. Bark thick and fibrous. Leaves scale-like, curving away from twig, 1cm long with tips usually raised; dark green with two bands of white dots underneath. Also called Giant Sequoia.
FLOWERS AND FRUITS Male flowers pale yellow; female flowers green, growing on tips of shoots. Cones, 7cm long, ripe in second year.
HABITAT Native only on the Pacific slopes of Sierra Nevada, California. Planted elsewhere. In Britain thrives best in west.

California Redwood
Sequoia sempervirens

Description Height usually 15–50m; can be over 100m. Evergreen and coniferous. Broadly columnar and fast-growing. Bark thick and spongy, scaling off to reveal red underneath. Young shoots green and hairless. On main branches, leaves scale-like all around stem, 6–8mm long; on branchlets, leaves linear in two rows to left and right of stem. Needles bright green above with two white bands underneath. Also called Coast Redwood.

Flowers and fruits Male flowers yellow on tips of small shoots; female flowers green, 6mm long. Egg-shaped cones, 2–2.5cm long, on tips of larger shoots.

Habitat Native to coastal moist areas of Oregon and California in North America. Ornamental in Europe.

Dawn Redwood
Metasequoia glyptostroboides

DESCRIPTION Height 15–30m. Deciduous. Trunk fluted with shaggy reddish bark. Sparse branches sweep upwards, producing a narrow conical tree. Leaves linear and flat, 1.5–2.5cm long.

FLOWERS AND FRUITS Flowers on shoots of previous year, in winter. Male cones rare. Female cones egg-shaped and rather pointed, about 2.5cm long, ripening to brown in first autumn. Buds unique, appearing below branches.

HABITAT Discovered in 1941 in China, and now grown in Europe as an ornamental tree. Grows fastest in wet areas.

Japanese Larch
Larix kaempferi

DESCRIPTION Height to 30m. Deciduous and coniferous. Branches long, thick and horizontal, although they can 'corkscrew'. Bark similar to that of European Larch (page 32) but more orange. Needles blue-green, singly on long shoots, clustered in tufts of about 40 on short shoots; purple cast in autumn.
FLOWERS AND FRUITS Male flowers yellow and globe-shaped; female flowers greenish or pink. Cones, 2.5–3cm long, squat and rounded with scales.
HABITAT Native to Japan. Widespread timber tree in north-west Europe.

European Larch
Larix decidua

DESCRIPTION Height 40m
or more. Deciduous and
coniferous. Lower stem
clear of branches on
mature tree. Branches
higher up sparse, thick
and horizontal. Bark light
brown. Needles light
green and soft, growing
singly on long shoots,
clustered in rosettes on
short shoots.

FLOWERS AND FRUITS Female
flowers loganberry-red
with green stripes; male
flowers yellow and globe-
shaped. Cones, 2–4cm
long, are egg-shaped.

ASSOCIATED SPECIES

**Common
Crossbill**

HABITAT Native to the Alps and Carpathians. Grows in mountains, but
now widely planted for forestry and as an ornamental. Loses leaves in
winter, which allows light to penetrate to the ground around it. This
encourages the growth of woodland spring flowers to a greater
extent than is the case with evergreen conifers. Seeds are a favourite
food of Common Crossbill – a species with a bill that is specially
adapted to pulling out seeds from pine cones.

Spanish Fir
Abies pinsapo

Description Height to 25m. Evergreen and coniferous. Regular conical shape with dull green foliage. Leaves arranged in almost perfect radial formation. Needles short, prickly, 1–2cm long, perpendicular to stem, hence the species' alternate name, Hedgehog Fir.

Flowers and fruits Male and female flowers grow on same tree. Male flowers large and cherry-red; female flowers pale green. Cones, 10–15cm long, with concealed bract scales, ripen to purplish-brown.

Habitat Native to south-west Spain, where it is rare. Introduced elsewhere. Prefers rocky chalky areas. In Britain sometimes grown as an ornamental.

European Silver Fir
Abies alba

DESCRIPTION Height 46m. Evergreen and coniferous. Narrow and conical shape. Crown cone-shaped with regular whorls of horizontal branches, upturned near their ends. Bark dark grey. Needles, up to 3.5cm long, shiny green on top and silvery beneath.

FLOWERS AND FRUITS Yellow male flowers grouped on undersides of twigs; green female flowers on upper sides near top of tree. Cones 10–15cm long.

HABITAT Native to central European highlands. Prefers cool and moist climate, and acidic soil.

Grand Fir
Abies grandis

Description Height to 100m. Evergreen and coniferous. Stout and fast growing with a narrow, symmetrical and conical crown. Young twigs olive-green with sparse minute hairs. Needles of various lengths, 2–6cm long, dark glossy green with two dark glossy green bands below; flattened and notched, spreading out to either side of shoot.
Flowers and fruits Cones, 8cm long, dark brown, erect, cylindrical, tapering at tips. Bracts concealed by cone scales.
Habitat Native to western North America. Planted for timber in wet areas of northern and central Europe.

Caucasian Fir
Abies nordmanniana

DESCRIPTION Height to 70m.
Evergreen and coniferous.
Stout trunk and conical
outline. Dark and densely
foliaged, retaining branches
almost to ground level, even in
old age. Young twigs sparsely
hairy. Dark green needles,
1.5–3.5cm long, flattened with
notched tips and two broad
white bands on undersides;
curve up and forwards, not
leaving a central parting
down shoot.

FLOWERS AND FRUITS Male
flowers reddish, underneath
shoots; female flowers greener
and upright, in separate
clusters on the same tree.
Cones, 10–20cm long,
cylindrical, with long and
deflexed bract beneath each
scale; ripen to dark brown.

HABITAT Native to mountain
areas of north-east Turkey and
Caucasus, but widely planted
in Europe for timber.

Douglas Fir
Pseudotsuga menziesii

DESCRIPTION Height 25–60m. Evergreen and coniferous. Tall and broadly conical. Bark dark grey. Needles flat, 2–3cm long, growing all around shoot, becoming dark glossy above, with two white bands beneath.
FLOWERS AND FRUITS Male flowers yellow; female flowers red, shaped like tassels. Cones, 7–10cm long, pendulous, light brown with three pointed bract scales projecting.
HABITAT Native to North America. Prefers moist and acidic sheltered conditions. Grown for timber in Europe.

Norway Spruce
Picea abies

DESCRIPTION Height 20–40m. Evergreen and coniferous. Regular conical shape. Bark light brown and smooth. Needles light green, short (1.4–2.5cm long) and prickly, growing on all sides of shoot.

FLOWERS AND FRUITS Male cones small, yellowish. Ripe female cones, 12–18cm long, brown, cigar-shaped, pendulous with rounded scales.

HABITAT Eastern France to south-east Russia, the Balkans up to Scandinavia. Grows in mountain forests, but more lowland in northern part of range. Widely planted for timber and Christmas trees. Pitch and turpentine were once made from the resin.

Sitka Spruce
Picea sitchensis

Description Height 20–60m. Evergreen and coniferous. Crown begins as a narrow cone, then broader with long and heavy branches. Needles slender, flat and sharply pointed, 1–2.5cm long, blue-green above with two white bands beneath.

Flowers and fruits Male flowers pale yellow; female flowers greenish-red. Cones, 5–10cm long, light brown. Seeds have thin papery wings.

Habitat Native to coastal western North America. Widely planted for timber throughout north-west Europe.

White Spruce
Picea glauca

DESCRIPTION Height 15–20m. Evergreen and coniferous. Slow-growing. Crown conical, becoming more spire-like with age. Pink-grey bark at first turns to ash-brown. Tapered leaves, 1–1.7cm long, bluish-green with silver-white lines.

FLOWERS AND FRUITS Flowers spring on shoots of previous season; male flowers are yellow. Fruit a long brown cone, 2.5–6cm long.

HABITAT Native to North America. Grown throughout northern Europe for timber.

Lodgepole Pine
Pinus contorta var. *latifolia*

Description Height 20–25m. Evergreen and coniferous. Crown cone-shaped. Bark scaly and brown. Leaves blue-green, tapered and usually twisted, 4–5cm long.

Flowers and fruits Flowers late spring on current season's growth. Male cones yellow. Fruit takes two years to mature to an egg-shaped brown cone, 5cm long.

Habitat Native to North America; coastal, from south-east Alaska to California. Grown in Europe for forestry and as an ornamental.

Maritime Pine
Pinus pinaster

DESCRIPTION Height 20–30m. Evergreen and coniferous. Crown cone-shaped, becoming domed. Bark orange-brown at first, then dark purple or rust-brown with deep fissures. Grey-green needles, 7–11cm long, in bundles of two.

FLOWERS AND FRUITS Flowers late spring on current season's growth. Male cones yellow-brown. Fruit matures in second autumn to shiny brown cone, 8–20cm long. Seeds 1cm with wings.

HABITAT Native to Atlantic France to Portugal, Mediterranean to Greece and Morocco. Grows on coastal sand dunes.

Stone Pine
Pinus pinea

Description Height 15–20m. Evergreen and coniferous. Crown conical at first, but branches radiate with age. Bark orange, fissured, developing deep furrows and scaly plates. Leaves grey-green, 8–18cm long, in bundles of two.

Flowers and fruits Flowers mid-summer on current season's growth. Male cones orange-brown. Fruit takes three years to mature to brown cone, 8–15cm long. Seeds 2cm long, edible.

Habitat Native to Mediterranean region. Grown on sandy sites in warm areas of Europe as an ornamental.

Western Yellow Pine
Pinus ponderosa

DESCRIPTION Height 20–40m. Evergreen and coniferous. Crown cone-shaped. Bark purple-grey to dark brown; old trees have deep fissures. Grey-green leaves in bundles of three, 11–22cm long. Also called Ponderosa Pine.

FLOWERS AND FRUITS Flowers in early summer on current season's growth. Male cones cylindrical, purple. Fruit matures in second year to a purple-brown cone, 6–16cm long, in autumn.

HABITAT Native to North America, growing on open hillside. Cultivated in Europe as an ornamental.

Weymouth Pine
Pinus strobus

DESCRIPTION Height 15–25m. Evergreen and coniferous. Irregular conical shape. Crown becomes domed with maturity. Bark dark grey, smooth when young, rough with maturity. Leaves grow in bunches at first, spreading in second year, grey-green, 8–10cm long.

FLOWERS AND FRUITS Flowers on current season's growth. Male flowers yellowish; female flowers pinkish, situated at ends of shoots. Clusters of male cones, slightly curved, 10–20cm long, appear at base of shoot.

HABITAT Native to North America, growing in mixed forest on sandy soils. Formerly widely planted for timber, but susceptible to blister rust and less common nowadays.

Corsican Pine
Pinus nigra

Description Height to 40m. Evergreen and coniferous. Crown cone-shaped at first, but becomes more like a column as branches spread. Needles grey-green, 12–18cm long, in bundles of two.

Flowers and fruits Flowers in early summer on current season's growth. Female flowers pink; male flowers yellow and purple. Fruit ripens in second autumn to a yellow or grey-brown cone, 5–9cm long.

Habitat Native to Sicily, Calabria and Corsica. In Britain planted for forestry, and in shelter belts and parks.

Scots Pine
Pinus sylvestris

DESCRIPTION Height to 36m. Evergreen and coniferous. Fully branched; narrow outline changes with maturity. Loses lower branches and forms a flatter spreading crown. Upper bark warm red, lower bark deeply fissured. Needles in pairs, twisted, blue-green, 5–7.5cm long.
FLOWERS AND FRUITS Female flowers crimson in pairs at the end of current year's growth; male flowers at base of shoot. Cones, 3–7cm long, mature when two years old. Seeds winged, and released when scales open.
HABITAT Native to highland Scotland, across northern Europe and Asia. Also Spain, Turkey and the Caucasus. Prefers sandy soil. Planted elsewhere. In Caledonian forests supports a unique assemblage of animals, including the endemic Scottish Crossbill, as well as Red Squirrels, Pine Martens, Capercaillies and Crested Tits.

ASSOCIATED SPECIES

Crested
Tit

Red
Squirrel

Japanese White Pine
Pinus parviflora

Description Height 20m. Evergreen and coniferous. Crown a broad cone shape, becoming more rounded or flat-topped. Bark smooth, purple-grey. Leaves mid-green, bundles of five, twisted, 2–6cm long.
Flowers and fruits Flowers in early summer on current season's growth. Female cones green or pink, male cones pink-purple. Fruit matures in second autumn to a sticky brown cone, 5–7cm long. Seeds 1–1.3cm long.
Habitat Native to Japan. Grows in temperate mixed forests. Cultivated as an ornamental in Europe.

Aleppo Pine
Pinus halepensis

DESCRIPTION Height 20m. Evergreen and coniferous. Conical when young, maturing to a rounded tree. Bark silver-grey, maturing to red-brown with age. Needles in pairs, radial along shoot, 6–11cm long; they stay on the tree for two years.

FLOWERS AND FRUITS Flowers on current season's growth, in early summer. Fruit matures in second autumn to a red-brown cone, 5–12cm long, but often stays on tree unopened for several years. Seeds have a wing 2cm long.

HABITAT Native to Mediterranean. Prefers dry hillside country. Often planted as a windbreak.

Arolla Pine
Pinus cembra

Description Height 25–40m. Evergreen and coniferous. Densely foliaged and retains even the lowest of its short and level branches. Bark scaly, marked with resin blisters. Twigs covered with brownish-orange hair. Needles 5–8cm long, shiny green, stiff and grouped in erect bundles of five crowded on twigs.

Flowers and fruits Cones, 5–8cm long, oval and short stalked, ripening from bluish to purple-brown over three years; cone scales rounded, thickened at tips and minutely hairy. Seeds are wingless.

Habitat Mountain species native to the Alps and Carpathians. Planted for timber in parts of northern Europe.

Mountain Pine
Pinus mugo

DESCRIPTION Height to 10m. Evergreen and coniferous. Shrubby upland pine that hugs the ground, seldom reaching its maximum height, though it can form a small conical tree. Numerous crooked spreading stems and branches. Paired needles, 3–8cm long, bright green, stiff and curved.

FLOWERS AND FRUITS Cones, 2–5cm long, in clusters of 1–3, oval, shiny brown, ripening in second year; exposed ends of scales usually flat with central boss bearing small spine.

HABITAT Native to high mountains of central Europe and the Balkan Peninsula. Often planted in northern Europe as a sand binder, and as a wind or avalanche break elsewhere.

Goat Willow
Salix caprea

Description Height 8–12m; may be up to 20m. Deciduous. Crown domed. Bark pale grey, becoming orange within fissures, smooth. Leaves egg-shaped, broader towards bases, dark to grey-green, 5–10cm long. Also called Pussy Willow and Sallow.

Flowers and fruits Flowers appear before leaves in spring. Female catkins green; male catkins silver-grey tinged with yellow, 3cm long. Fruit ripens May–June, with many woolly seeds on female catkins ('pussy willows'), 3–7cm long.

Habitat Common and widespread in northern Europe except south-west. Grows in damp and dry woods and scrub; also grown as a garden tree.

Grey Willow
Salix cinerea

DESCRIPTION Height 5–15m. Deciduous. Crown egg-shaped. May grow on several stems. Bark starts off smooth, greyish-brown, and turns brown with fissures. Leaves oval to lance-shaped, uppersides hairy at first, dark shiny green, grey-blue underneath, 2–9cm long.

FLOWERS AND FRUITS Flowers early spring, before leaves appear. Female catkins green; male catkins pale yellow, 2–3cm long. Fruit capsules contain silky seeds.

HABITAT Damp areas and hedgerows throughout western Europe.

Crack Willow
Salix fragilis

DESCRIPTION Height 10–20m. Deciduous. Crown cone-shaped when young, becoming broad and rounded in mature trees. Bark starts off scaly, becomes deeply fissured, grey. Leaves lance-shaped, shiny green uppersides, bluish-white underneath, 9–15cm long. Twigs readily break (hence its scientific name, *fragilis*) and root easily.

FLOWERS AND FRUITS Flowers April–May. Trees are either male or female. Male catkins pale yellow, 4–6cm long; female catkins green. Woolly white, usually sterile seeds formed on female catkins.

HABITAT Native to central and southern Europe, including Britain, and widespread. Grows in damp lowland woodland, and along the banks of rivers and canals. Hybridises freely with other willow species. Willows are a preferred foodplant of the striking larvae of Puss Moths.

ASSOCIATED SPECIES

Puss Moth

Puss Moth
larva

White Willow
Salix alba

DESCRIPTION Height 10–30m. Deciduous. Mature crown domed with ascending branches. Bark shows deep fissures and is grey-brown. Leaves lance-shaped, shiny with hairs, grey-green uppersides, almost white beneath, 5–10cm long.

FLOWERS AND FRUITS Flowers April. Male and female trees are separate. Pale yellow catkins; female catkins 3–4cm long, male catkins 4–5cm long. Fluffy seeds, formed on female catkins, ripen in July.

HABITAT Europe to central Asia. Favours damp soil, particularly near rivers and ponds.

Weeping Willow
Salix x sepulcralis

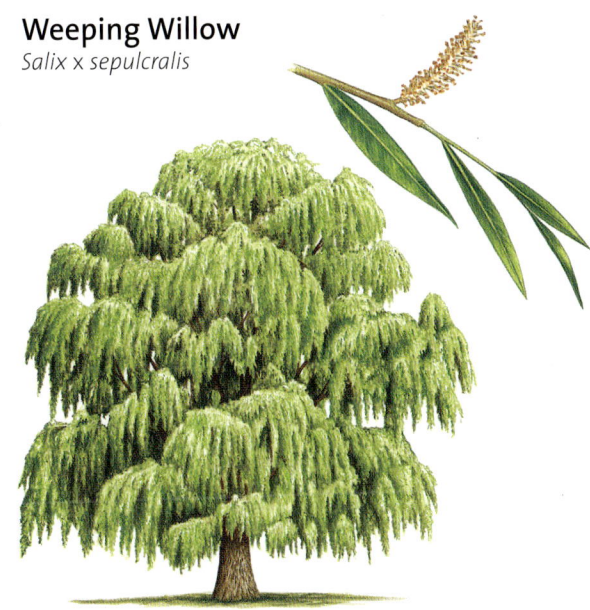

Description Height 10–18m. Deciduous. Crown rounded with branches growing downwards. Bark golden-green when young, smooth; becomes grey-brown and develops deep fissures. Leaves lance-shaped with tapered upper surfaces having white hairs at first, blue-green beneath, 7–13cm long.

Flowers and fruits Flowers in spring on short leafy shoots. Usually only male catkins appear, which curve upwards.

Habitat Widely cultivated hybrid between *S. alba* and *S. babylonica*. Associated with water and large gardens.

Common Osier
Salix viminalis

DESCRIPTION Height 10m. Deciduous. Narrow upright crown and grey bark with fissures. Leaves narrow, lance-shaped and tapering, with dull green upsides and silver-grey beneath, 10–25cm long.
FLOWERS AND FRUITS Flowers appear before leaves. Male and female flowers grow on separate twigs. Catkins are erect, and male flowers have yellow anthers. Fruit is flask-shaped.
HABITAT Native to central and eastern Europe, and found in Britain. Prefers damp places. Cultivated for basket weaving.

Bay Willow
Salix pentandra

DESCRIPTION Height to 20m. Deciduous. Highly glossy leaves and shiny twigs give a varnished appearance to this small tree or small shrub. Branches spreading with reddish-brown twigs. Leaves elliptical to lance shaped, long-pointed, leathery, dark and very shiny above, paler beneath, 5–12cm long.

FLOWERS AND FRUITS Hairy-stalked catkins appear with leaves May–June. Dense cylindrical male catkins, 2–5cm long, pale yellow and borne on separate trees from shorter greenish female catkins. Seeds silky and plumed.

HABITAT Common along waterways and in wet soils in most of Europe except the Mediterranean islands.

White Poplar
Populus alba

DESCRIPTION Height to 25m. Deciduous. Tall, often leaning in the soft ground it favours. Crown columnar, open and with twisting branches. Bark grey-white on upper tree. Young twigs densely white and hairy. Leaves have five small pointed lobes, white underneath, 6–12cm long.

FLOWERS AND FRUITS Male and female flowers appear February–March on separate trees, before leaves emerge. Female flowers are green; male flowers are red. Fruits consist of fluffy seeds.

HABITAT Native in western Europe to central Asia, and thought to be an early introduction to Britain. Grows in wet areas, near roads and on sea coasts. Wind-resistant and suckers freely, so often grown near to the sea to stabilize loose sand. Slower growing than Black Poplar (page 65). Poplars are the preferred foodplants of Poplar Hawkmoths.

ASSOCIATED SPECIES

Poplar
Hawkmoth

Poplar
Hawkmoth
larva

Grey Poplar
Populus x canescens

DESCRIPTION Height 25–40m. Deciduous. Crown columnar and open. Young bark pale, whitish; lower trunk black. Leaves placed alternately and vary considerably in shape, white undersides. Suckers readily from roots.

FLOWERS AND FRUITS Male catkins are green; female catkins have a golden tinge. They grow on separate trees.

HABITAT Native to central Europe, growing on damp ground, near rivers and in water meadows. Probably introduced to Britain very early.

Black Poplar
Populus nigra

Description Height to 35m. Deciduous. Crown a large dome with spreading branches. Bark grey-brown with fissures and sometimes burrs. Leaves oval with long pointed tips, about 8cm long.

Flowers and fruits Male red and female green catkins form on separate trees. Fluffy seeds disperse in June.

Habitat Probably native to Britain. Widely planted as a windbreak and beside roads, and flourishes by water. Tolerant of pollution, so sometimes planted in cities.

Balm of Gilead
Populus candicans

DESCRIPTION Height to 25m.
Deciduous. Broad,
spreading suckering tree
with a strong balsam
scent, especially on wet
spring days as buds break.
Grey-green bark. Young
twigs fragrant, thinly
hairy, soon becoming
glossy brown. Leaves
fragrant, heart-shaped,
downy beneath, 5–15cm
long. Leaf stalk slightly
flattened on upper and
lower sides, with two
glands near the top.
FLOWERS AND FRUITS Catkins
appear in March, just
before leaves. Only
females known; 4–6cm
long in flower, reaching
16cm before releasing
silky hairy seeds.
HABITAT Native to North
America. Commonly
planted in Europe, and
often naturalized.

Aspen
Populus tremula

DESCRIPTION Height 15–25m. Deciduous. Crown cone- or column-shaped. Suckers easily, often growing in small groves of the same sex. Bark green-grey, smooth. Leaves round to oval with sharp tips, toothed; uppersides copper coloured at first, becoming blue-green, and yellow in autumn, 1.5–8cm long. Leaves rustle in the lightest wind.

FLOWERS AND FRUITS Male and female flowers grow on separate plants. Male flowers egg shaped, green and brown; female flowers green. Seeds in 4cm-long catkins in May.

HABITAT Found from Europe to Asia Minor and central Asia; native to Britain. Favours damp hillsides and woods.

Common Walnut
Juglans regia

DESCRIPTION Height 15–25m. Deciduous. Crown rounded and branches radiating. Bark smooth, shiny and grey at first, becoming ridged with wide fissures. Leaves leathery, aromatic, pinnate with 5–9 oval leaflets, 20cm long.

FLOWERS AND FRUITS Flowers May–June. Male catkins 5–10cm long, appearing at ends of new shoots; female catkins in clusters on old wood. Globe-shaped fruits have a thick green fleshy case, splitting to reveal a large nut, 4–5cm long, which has an excellent flavour.

HABITAT Native to parts of Asia and south-east Europe. Grows in woodland. Planted and naturalized since ancient times across Europe. Foodplant of the larvae of some moth species, including Emperor Moths and Brown-tails. Nuts are eaten by mice and squirrels, and by large birds such as Rooks.

ASSOCIATED SPECIES

Rook

Emperor
Moth

Emperor
Moth larva

Black Walnut
Juglans nigra

DESCRIPTION Height 15–35m. Deciduous. Crown domed or rounded. Bark usually dark brown, deeply furrowed and ridged. Leaves large, 30–60cm long, with 9–23 lance-shaped tapered leaflets.

FLOWERS AND FRUITS Small green flowers appear in spring with yellow male catkins. Fruit smooth with a thick fleshy covering over a large nut, 3–5cm long. Edible and good.

HABITAT Native to North America. Also grown as an ornamental elsewhere. Prefers streamside sites. The husk can be used to make a dark dye.

Grey Alder
Alnus incana

DESCRIPTION Height 20m. Deciduous. Smooth grey-green bark with distinct pores. Leaves alternate, pointed, elliptic, 5–10cm long. Upper surface dull green; lower surface bluish-green and hairy.

FLOWERS AND FRUITS Separate male and female catkins on the same tree. Male catkins long and drooping; female catkins short, egg-shaped and turning to green fruits, 1.5cm long, which remain on tree all winter, maturing in autumn to dark woody cones.

HABITAT Native to northern and central Europe, favouring mountain valleys and sub-arctic forests. Introduced to Britain.

Common Alder
Alnus glutinosa

DESCRIPTION Height 15–25m. Deciduous. Broad domed or conical crown with regular branches. Bark rough and often sprouts shoots. Leaves rounded and alternate, sometimes notched at tips, 4–10 long. Nodules on roots contain nitrogen-fixing bacteria.

FLOWERS AND FRUITS Male and female catkins grow on the same tree. Buds are stacked with green fruits, ripening in spring. Female cones lengthen to 1–1.5cm, becoming black and woody, and stay on tree all winter, opening to shed nutlets.

HABITAT Native to Europe except far north. Thrives in wet ground, lining banks of rivers and streams. Seeds are a key source of winter food for finches such as Lesser Redpolls and Siskins. Leaves are the foodplant of many insects, particularly moth larvae. The wood was once used for clogs, and as a source of charcoal for gunpowder.

ASSOCIATED SPECIES

Lesser
Redpoll

Siskin

Green Alder
Alnus viridis

DESCRIPTION Height 5m. Deciduous. Small tree or large shrub. Bark brown. Leaves more pointed than those of Common Alder (page 72) and with sharply double-toothed margins; sticky to the touch when first open, 4–9cm long.

FLOWERS AND FRUITS Male catkins appear with leaves, yellow and pendulous, to 5–12cm long; female catkins, 1cm long, in clusters of 3–5, erect and greenish at first, turning reddish. Rounded, green and tough ripe catkins become blackish and persist until following spring; contain broadly winged nutlets.

HABITAT Native to mountains in central and eastern Europe, north to Sweden. Planted in Britain as an ornamental.

Downy Birch
Betula pubescens

DESCRIPTION Height 8–14m. Deciduous. Branches twist, seldom hang down and form a round-headed tree. Bark red-brown. Leaves rounded, hairy on undersides, margins toothed, 1.5–5.5cm long.

FLOWERS AND FRUITS Male and female flowers grow in separate catkins on same tree, opening in April. Male catkins pendulous; female catkins smaller and more erect. Fruits stay on tree in winter, breaking up into winged windborne seeds.

HABITAT Found across Europe to central Asia in moist areas and mountain forests. Like Silver Birch (page 76), associated with many fungi, as well as insects, which feed on its leaves.

Silver Birch
Betula pendula

DESCRIPTION Height 14–30m. Deciduous. Young bark reddish maturing to black marked with silver-white. Slender habit, with pendulous branches forming a pointed crown when young, domed when mature. Leaves alternate on slender hairless stalks, triangular, pointed, margins with small teeth, 3–7cm long.

FLOWERS AND FRUITS Male and female flowers in separate drooping catkins on same tree in April–May. Inconspicuous and wind-pollinated. Fruiting catkins stay on tree until winter, then break up into scaled and winged windborne seeds.

HABITAT Found throughout Europe to northern Asia, preferring light dry soils. One of Britain's native trees. Base often rich in fungi, with a number of species closely associated, including Brown Birch Boletus, Coconut-scented Milk Cap, Birch Polypore, Fly Agaric and Woolly Milk Cap.

ASSOCIATED SPECIES

Brown Birch Boletus

Coconut-scented Milk Cap

Himalayan Birch
Betula utilis

DESCRIPTION Height 10–20m. Deciduous. Bark red-brown and white. Pendulous branches form a pointed crown when young, a dome when mature. Leaves egg-shaped, tapered, singly or in pairs on spur shoots, alternate on long shoots, toothed, 3–10cm long.

FLOWERS AND FRUITS Male and female flowers on separate catkins on the same tree in spring. Male catkins yellow and pendulous; female catkins erect. Fruiting catkins release winged seeds.

HABITAT Native from the Himalayas to Tibet and western China. Widespread as an ornamental.

Paper-bark Birch
Betula papyrifera

DESCRIPTION Height 10–20m. Deciduous. Bark warm brown, peeling to expose white underneath; continually renewed by further peeling. Leaves egg-shaped, tapered to a point, toothed, uppersides matt green, light green and shiny beneath, 5–10cm long. Also called Canoe Birch.

FLOWERS AND FRUITS Produces pendent yellow male catkins, and pendent cylindrical fruits that release winged seeds.

HABITAT Native to North America. Grows in large forests in moist sites in the wild. Grown as an ornamental elsewhere. Bark was once widely used to make canoes, baskets and covers for wigwams.

Common Hazel
Corylus avellana

Female flower

Male catkin and seeds

DESCRIPTION Height 8–10m. Deciduous. Rounded crown. Bark light brown, becoming grey-brown in older trees. Leaves rounded, heart-shaped at bases, downy, serrated, 5–10cm long. Yellow in autumn.
FLOWERS AND FRUITS Flowers late winter on previous season's growth. Female flowers from buds; male catkins yellow and pendulous. Fruit a brown nut, 1.5–2cm long, with a leafy husk.
HABITAT Native to Britain and other parts of Europe. Occurs naturally in woodland understorey. Grown for its nuts. Nuts eaten by squirrels, voles and mice, as well as Nuthatches and other birds.

Hornbeam
Carpinus betulus

DESCRIPTION Height 15–25m. Deciduous. Rounded crown in mature trees. Bark smooth, fluted in old trees, silver-grey in colour. Leaves broadly elliptical, acutely tapering, matt green, toothed, 4–5cm long. Yellow and gold in autumn.

FLOWERS AND FRUITS Flowers in spring. Female flowers green; male flowers drooping catkins, 3–5cm long. Fruit a leafy bract, set in pairs. Seeds flat and egg-shaped.

HABITAT Widespread across Europe to Asia Minor. Grows in woodland on heavy soils. Tough seeds are an important food for birds, particularly Hawfinches, which also favour woodland containing Hornbeams for nesting.

ASSOCIATED SPECIES

Hawfinch

Common Beech
Fagus sylvatica

DESCRIPTION Height 20–35m. Deciduous. Domed crown, more conical in young trees than in older individuals. Bark silver-grey, smooth. Leaves oval, toothed, light green becoming darker, 6–10cm long. Yellow and brown in autumn.

FLOWERS AND FRUITS Female flowers green; male flowers globe-shaped on drooping stalks. Fruit is a small, slightly prickly woody cup containing small nuts.

HABITAT Native to western and southern Europe, including southern England and southern Sweden. Widely planted. Grows in woodland on free-draining acidic and alkaline soils. In winter beech nuts lying on the ground (beechmast) are a favourite food of finches like Bramblings and Chaffinches, as well as small mammals. Many fungi grow at the bases of the trees in autumn, including Horn of Plenty, Amethyst Deceiver, Porcelain Fungus, Artist's Bracket and Death Cap.

ASSOCIATED SPECIES

Bramblings (left and above) and Chaffinches (top)

Sweet Chestnut
Castanea sativa

DESCRIPTION Height 20–30m. Deciduous. Crown conical in young trees; older trees have a broad-domed or spreading habit. Bark grey, smooth at first, fissured later, often spiralling up trunk. Leaves lance-shaped, 15–20cm long. Yellow in autumn.

FLOWERS AND FRUITS Flowers mid-summer. Male catkins erect, creamy white, long and pendulous. Female flowers greenish. Fruit has a large, round prickly green case containing 1–3 nuts. Nuts are edible and good.

HABITAT Native to Mediterranean region and widely planted elsewhere. Prefers acidic soils.

Turkey Oak
Quercus cerris

DESCRIPTION Height 20–40m. Deciduous. Fast-growing oak. Crown is cone-shaped at first, becoming rounded, then domed. Bark dull grey to silver-grey, fissured. Leaves oval with variable lobing, 5–14cm long. Brown in autumn.

FLOWERS AND FRUITS Flowers in early summer. Inconspicuous female flowers. Male catkins drooping, 5–6cm long, crimson turning yellow-brown. Fruit is a plump acorn in a scaly cup.

HABITAT Native to southern Europe, growing in woodland. Widely planted throughout Europe.

Pyrenean Oak
Quercus pyrenaica

DESCRIPTION Height 20–25m. Deciduous. Rather open crown is dome-shaped. Bark light grey with deep fissures breaking into small square scales. Leaves oval, broader towards tips and lobed, 10–20cm long.

FLOWERS AND FRUITS Flowers late June with golden pendent male flowers. Acorn matures in one year; it is oblong and has a neat hemispherical cup holding the lower third.

HABITAT Native to woodland in north-west France, Portugal, Spain and Morocco. Occasionally planted in Britain.

Holm Oak
Quercus ilex

DESCRIPTION Height 30m. One of the hardiest evergreen oaks. Young tree grows like a column, crown later becoming domed. Bark brown-black, smooth at first then cracking into small squares. Leaves usually oval or lance-shaped, 4–8cm long; shed in summer of their second year.

FLOWERS AND FRUITS Flowers in early summer; male catkins yellow, 4–7cm long. Acorn 1.5–2cm long in clusters of 1–3.

HABITAT Native to Mediterranean. Prefers dry woods and coastal cliffs. Introduced to Britain, where it is naturalized in milder southern areas. The most frequently encountered evergreen oak in northern Europe.

Sessile Oak
Quercus petraea

ASSOCIATED SPECIES

DESCRIPTION Height 20–40m. Deciduous. Taller than Pedunculate Oak, with which it hybridizes. Bark grey-brown, smooth at first, later fissured and cracked. Leaves lobed, 8–14cm long.
FLOWERS AND FRUITS Flowers late spring: female flowers in groups of 2–6 towards tips of current season's shoots; male flowers pendent from buds on ends of previous year's shoots, 5–8cm long. Acorns without stalks.
HABITAT Grows in much of Europe, including Britain. Prefers light acid soils. Often rich in epiphytes, and associated with hole-nesting birds such as flycatchers, which feed on the abundant insect life that the tree supports.

Pied Flycatcher

Pedunculate Oak
Quercus robur

Description Height 15–45m. Deciduous. Young trees have a cone-shaped crown, which is wide-domed in open situations. Mature trees massive, often dominating woodland. Individuals can live for more than 1,000 years. Bark initially smooth and grey-green, then fissured. Leaves 4–12cm long, with 4–6 lobes. Also called English Oak.

Flowers and fruits Flowers in late spring: male flowers hang in long catkins; female flowers are much less conspicuous. Acorns in stalked clusters of 1–4 in shallow cups.

Habitat Widespread across Europe. Grows in woodland, hedgerows and open parkland, on heavy alkaline soils. Mature individuals provide habitats for thousands of animals (see page 90). Wood was once used exclusively in medieval timber-framed buildings and for shipbuilding.

The Pendunculate Oak as a Habitat

A mature oak provides a habitat for thousands of living creatures, including more insect species than any other European plant. Among them are the Purple Hairstreak, the only British butterfly dependent on the tree for its larva's food; the Purple Emperor, the males of which establish territories on prominent trees; the larvae of the Green Oak Tortrix Moth, among the most abundant insects on oaks; the Oak Apple Gall, caused by a wasp that lays its eggs in oak rootlets that then become galls, each gall containing a single larva; and the Oak Bush-cricket, the only true tree-dwelling bush-cricket.

About 40 hectares of oakwood can support 300–400 bird species in summer, which rely on the abundance of food and nest sites. Most are small insect-eating passerines that benefit from the abundance of insects at the time when they are nesting. Jays collect and bury acorns, often quite a way from the parent oak, thus helping in their dispersal. In autumn oakwoods produce rich crops of fungi such as Summer Boletus, Beefsteak Fungus and Oak Milk Cap.

Beefsteak Fungus

Summer Boletus

Purple
Emperor

♀   ♂

Purple
Hairstreak

♀

Green Oak
Tortrix Moth

Oak Bush-cricket

Oak Apple
gall

Oak Apple
Gall wasp

Jay

Pin Oak
Quercus palustris

DESCRIPTION Height 20–25m. Deciduous. Crown cone-shaped, slender at first, domed in old trees. Bark smooth, dark silver-grey, fissured and darker later. Leaves round or oval, 7–13cm long, with 2–3 pairs of lobes. Vivid red autumn colour.

FLOWERS AND FRUITS Female flowers in axils of new growth in groups of 3–4. Round acorns 1.2cm long ripen in second autumn.

HABITAT Native to North America, growing in wet sites in the wild. Widely cultivated in Europe.

Red Oak
Quercus rubra

DESCRIPTION Height 20–35m. Deciduous. Crown cone-shaped, becoming wide-domed on radiating branches. Leaves, 7–20cm long, have 3–5 pairs of large toothed lobes. Autumn colour is red, sometimes yellow or brown.

FLOWERS AND FRUITS Flowers May: female flowers in axils of new growth; male catkins. Acorns in broad shallow cups, 1.5–2.8cm long.

HABITAT Native to eastern North America, and widely cultivated. In Europe grown as an ornamental and for forestry.

Downy Oak
Quercus pubescens

DESCRIPTION Height to 25m. Deciduous. Small tree or even a shrub, but capable of reaching a moderate height. Twigs densely grey-hairy. Leaves, 4–12cm long, have shallower and more forwards-pointing lobes than those of other deciduous oaks; undersides of leaves grey, velvety, gradually becoming smooth.

FLOWERS AND FRUITS Flowers May: males in catkins; females in clusters of 1–3. Acorns in stalkless shallow cups with close-pressed and grey-woolly scales, 1.5cm long. Ripen in same year.

HABITAT Native to south-central and western Europe, growing on dry limestone hills. Occasionally planted in Britain.

Kermes Oak
Quercus coccifera

Description Height to 10m.
Evergreen. Small tree or
shrub resembling a
Holly bush. Young twigs
yellowish with star-
shaped hairs. Leaves
oval to oblong, 1.5–4cm
long, sometimes heart-
shaped at bases,
with wavy and spine-
toothed margins. Leaf
stalks only 1–4mm long.
Flowers and fruits
Flowers April–May: males
in catkins; females in
clusters. Shallow acorn
cups with stiff, more or
less spiny scales. Acorns
ripen in second year.
Habitat Widespread
across Mediterranean,
especially in hotter and
drier parts, and has long
been grown in northern
European collections.

Wych Elm
Ulmus glabra

DESCRIPTION Height 20–30m. Deciduous. Crown in young trees egg-shaped, becoming broadly domed. Bark smooth, silver-grey, turning grey-brown and developing fissures. Leaves in sprays, oval, 8–18cm long. Leaves turn yellow in autumn.

FLOWERS AND FRUITS Reddish-purple flowers appear before leaves. Fruit disc-shaped with seed in centre surrounded by circular wing; fruits hang in bundles.

HABITAT Occurs throughout Europe except far north. Grows in woodland and hedgerows.

English Elm
Ulmus procera

DESCRIPTION Height to 30m. Deciduous. Crown egg- or cone-shaped at first, becoming a tall dome. Suckers grow from roots. Leaves rounded to oval, toothed, dark green, rough, 4–10cm long.

FLOWERS AND FRUITS Flowers towards end of winter/early spring. Small reddish flowers are produced on previous season's shoots. Fruit disc-shaped with seed set in centre of circular wing, 1.2cm long.

HABITAT Southern England, in hedgerows and fields. Also western and central Europe. Proliferated during enclosure, but decimated in latter half of 20th century due to Dutch elm disease, a fungal growth spread by a small beetle. Timber is valuable.

European White Elm
Ulmus laevis

DESCRIPTION Height to 35m. Deciduous. Tall and open tree with wide-spreading crown. Bark initially smooth, becoming deeply ridged with age. Leaves oval to nearly round, 6–13cm long, usually smooth above but downy grey beneath, with 12–19 pairs of veins and very asymmetric leaf bases.

FLOWERS AND FRUITS Flowers March in reddish clusters, before leaves appear. Fruit pendulous with a fringe of white hairs on papery wings, 1–1.2cm long.

HABITAT Native to central and south-eastern Europe, mainly in river valleys. Occasionally planted for shelter elsewhere. Very rare in Britain as an old wild tree (which may be native).

Smooth-leaved Elm
Ulmus minor

Description Height 20–30m.
Deciduous. Crown cone-
shaped, domed in older trees.
Bark smooth, silver-grey,
developing deep fissures. Leaves oval, 6–8cm long, base with one
rounded side, one wedge-shaped, toothed, shiny green.

Flowers and fruits Purple-red flowers appear early spring on previous
season's growth. Fruit a round flat disc with a seed in the centre
surrounded by a thin wing, 1.2–1.5cm long.

Habitat Native to western Europe and North Africa to south-west
Asia. A tree of lowland woodland.

Southern Nettle-tree
Celtis australis

DESCRIPTION Height to 14m.
Deciduous. Leaves resemble
those of nettles, but do not
sting; narrowly oval and
sharply toothed, 4–15cm
long, tapering to an often twisted lip, with the upper surface stiffly
hairy, lower downy-white, especially on the veins.

FLOWERS AND FRUITS Flowers appear with young leaves in May; solitary
in leaf axils; dull brownish-green. Berry-like fruit, fleshy and edible,
long-stalked, ripening from brownish-red to black, 1cm long.

HABITAT Native to southern Europe, where it can live to 1,000 years, and
planted elsewhere as a street tree and ornamental. Rare in Britain.

Fig
Ficus carica

DESCRIPTION Height 10–15m. Deciduous. Crown low, broad and spreading. Pale grey bark has fine wrinkles. Leaves have 3–5 large rounded lobes, and are dark green, 30cm long.

FLOWERS AND FRUITS Flowers pear-shaped and tiny, largely hidden within shoot tips. Pollination is by insects. Fruit green maturing to purple, pear-shaped with many seeds enclosed within flesh, to 5cm long.

HABITAT Originated in western Asia and now widespread in Mediterranean. Widely grown for its delicious fruits. Long naturalized in Britain.

Black Mulberry
Morus nigra

DESCRIPTION Height to 10m. Deciduous. Crown a low dome. Bark orange-brown, brighter in fissures. Leaves broad, egg-shaped, wider towards bases, pointed tips, dull-green, rough, 8–12cm long.
FLOWERS AND FRUITS Inconspicuous flowers appear in May in leaf axis. Fruit starts green, turning purple-black, 2–2.5cm long. Fruits edible and good, but usually used to flavour dishes made with other fruits.
HABITAT Origin uncertain, but probably Asiatic. Widely cultivated. In Britain grown in gardens and parks in warmer areas; rarely naturalizes. Grown both as a fruit tree and as an ornamental.

Tulip Tree
Liriodendron tulipifera

DESCRIPTION Height 15–25m. Deciduous. Bark develops rough ridges. Twigs have alternate long-stalked hairless leaves, 7–12cm long, which develop into four lobes with flattened tops.

FLOWERS AND FRUITS Flowers June–July, tulip-shaped, 4–5cm long with three spreading greenish-white sepals and six erect petals that are pale green with a broad orange band near their base. Fruits form in dense cone-like aggregate, each fruit breaking into two dry, long-winged seeds.

HABITAT Native to North America. Grown as an ornamental tree elsewhere.

Southern Evergreen Magnolia
Magnolia grandiflora

DESCRIPTION Height to 30m. Evergreen. Spreading tree with very large and conspicuous flowers, its large branches forming a conical crown. Leaves thick and leathery, 8–16cm long, very shiny above, covered with rusty hairs beneath.

FLOWERS AND FRUITS Flowers July–November. Blooms up to 15cm diameter with six petal-like segments; initially conical, gradually opening almost flat. Fruit forms a cone-like structure, 5–6cm long.

HABITAT Native to eastern North America. Widely grown in Europe as an ornamental, as are several other magnolias.

Bay
Laurus nobilis

DESCRIPTION Height to 20m. Evergreen. Vigorously suckering shrub or small tree that is densely branched and pyramidal in shape with dark blackish bark. Hairless lance-shaped leaves, 5–12cm long, with numerous oil glands. Strongly aromatic. Also called Laurel, Poet's Laurel and Sweet Bay.

FLOWERS AND FRUITS Male and female flowers grow on separate trees. Female flower has a single ovary ripening to a single-seeded black berry, 1.2cm long.

HABITAT Native to Mediterranean, growing in thickets and forest. Widely planted elsewhere. Often planted in British gardens and parks; rarely naturalizes.

Witch Hazel
Hamamelis virginiana

DESCRIPTION Height to 5m. Deciduous. Shrub or small tree with upwards-pointing branches. Leaves alternate with small teeth, unequal bases and long pointed tips, 10–15cm long, deep green and shiny uppersides. Leaves turn yellow in autumn.

FLOWERS AND FRUITS Flowers in autumn before leaves fall. Fruit a 1cm-long capsule enclosing two black seeds.

HABITAT Native to eastern North America. Grown in Europe as an ornamental and for its medicinal properties. In Britain sometimes naturalized in open woodland. Distilled from leaves and twigs to make a well-known skin lotion.

London Plane
Platanus x hispanica

DESCRIPTION Height 20–40m. Deciduous. Hardy natural hybrid with a domed crown and thick twisting branches. Bark dark and flakes in autumn, showing creamy patches. Leaves alternate with five lobes, to 24cm long.

FLOWERS AND FRUITS Female flowers reddish, appearing at shoot tips; male flowers yellow and situated further back on stems. Bobble-like balls of hairy seeds remain on tree during winter, breaking up in spring.

HABITAT Origin unknown, but probably noted in continental Europe. Widespread as an ornamental tree, often lining town avenues due to its tolerance of pollution.

Oriental Plane
Platanus orientalis

DESCRIPTION Height 15–30m. Deciduous. Very long-lived hybrid. More rounded crown than that of London Plane (page 107). Bark pinkish-brown and flaky. Leaves to 18cm long, with 5–7 lobes.

FLOWERS AND FRUITS Female flowers reddish, appearing at shoot tips; male flowers yellow and situated further back on stems. Bobble-like fruits remain on the tree during winter.

HABITAT Native of south-east Europe and Asia Minor, growing near rivers and streams. Long grown in Britain in warmer areas.

Orchard Apple
Malus domestica

DESCRIPTION Height to 15m. Deciduous. Crown dome-shaped or rounded. Bark grey-brown and smooth. Leaves oval, slightly wider towards bases, toothed, dull green, to 13cm long.

FLOWERS AND FRUITS Flowers April–May in clusters of 4–7 flowers, which are pinkish or white. Fruit is the well-known apple, 5–10cm long, green, russet or red in colour, depending on variety, and edible.

HABITAT Originated in Asia. Widespread in temperate regions as a cultivated tree for its fruits.

Wild Pear
Pyrus pyraster

DESCRIPTION Height 12–25m. Deciduous. Cone-shaped crown in young trees, becoming domed in older individuals. Bark brown to blackish with small square plates. Leaves rounded with short pointed tips, to 7cm long; leathery shiny green uppersides. Leaves flutter and rustle in the lightest wind.

FLOWERS AND FRUITS Flowers appear before leaves; white clusters, outer flowers open first. Fruit up to 4cm long, yellow-green and hard.

HABITAT A wild European pear, growing in hedgerows and woods. Often confused with Common Pear (*P. communis*).

Medlar
Mespilus germanica

DESCRIPTION Height 5–7m. Deciduous. Low crown and spreading branches. Bark grey-brown with fissures and oblong plates. Leaves lance-shaped or oval, 5–15cm long, with dark yellow-green uppersides, tiny teeth around margins.

FLOWERS AND FRUITS Flowers May–June. White flowers occur singly on terminal shoots. Fruit pear-shaped and edible, 2–3cm long; not usually picked until after it has been frosted.

HABITAT Native to south-east Europe to Iran. Naturalized in central Europe. Grows in woodland. Long grown in Britain; sometimes naturalized in south-east England.

Wild Crab
Malus sylvestris

DESCRIPTION Height 2–9m. Deciduous. Small and spreading with large twisted branches that curve downwards, and dense foliage. Bark dark brown with fissures and cracks. Leaves oval to egg-shaped, wider near tips, 3–7cm long, with serrated edges.

FLOWERS AND FRUITS Flowers have white or pink petals. Fruit globe-shaped, like a small apple, and greenish. It is edible but sour, and is usually used to make jelly.

HABITAT Native to chalky hilly regions in much of Europe, and domesticated in ancient times. Widespread, favouring old woodland and hedgerows. One of the parent species of cultivated apples. The colourful flowers attract bees in spring. Fruits are eaten by birds such as Robins and Starlings in winter.

ASSOCIATED SPECIES

Robin

Starling

Common Rowan
Sorbus aucuparia

DESCRIPTION Height 5–20m. Deciduous. Cone-shaped crown in young trees, becoming rounded or spreading in older individuals. Bark smooth, shiny and grey. Older trees grey-brown with ridges. Leaves pinnate with about 13–15 leaflets, each to 6m long, with serrated margins. May turn bright yellow or red in autumn. Also called Mountain Ash.

FLOWERS AND FRUITS Flowers late spring with large clusters of creamy-white flowers. Fruits are red berries, usually ripened by August.

HABITAT Native to Europe, the Caucasus and North Africa. Planted as an ornamental. Bright red fruits are liked by birds in winter, even attracting migrants such as Waxwings to busy town centres.

ASSOCIATED SPECIES

Waxwing

♀

♂

Swedish Whitebeam
Sorbus intermedia

DESCRIPTION Height
to 15m. Deciduous.
Rather squat and
rounded medium-
sized tree with a
short trunk and
a domed crown.
Leaves glossy green, 6–12cm
long, seven pairs of veins, felted
beneath with yellowish-grey
hairs. Lobes near base of blade
reach one-third of the way to
midrib; those towards tip progressively shallower, eventually
reduced to coarse teeth.

FLOWERS AND FRUITS Flower clusters in May; white five-petalled flowers,
1.2–2cm diameter. Fruit oblong-ovoid, scarlet with a few warts,
1.2–1.5cm long.

HABITAT Native to Scandinavia and the Baltic region. Tolerant of air
pollution and commonly used as a street tree in Britain.

Common Whitebeam
Sorbus aria

Description Height 15–25m. Deciduous. Young trees have a cone-shaped crown that becomes domed in older trees. Bark grey, smooth at first, turning scaly and fissured. Leaves egg-shaped, margins serrated, 6–12cm long, with white undersides.

Flowers and fruits Dull white flowers appear in clusters May–June. Fruits globe-shaped, bright red, in hanging clusters.

Habitat Grows in southern England through southern Germany, Italy, Spain, Morocco and the northern Balkans. Prefers limey or chalky soils, but tolerates sandy conditions. Favours well-drained open woodland.

True Service
Sorbus domestica

DESCRIPTION Height to 20m. Deciduous. Long-lived. Branches spread to horizontal. Bark shredding. Leaves pinnately divided into 6–8 pairs of oblong leaflets; each leaflet 3–5.5cm long, blade symmetrical at base, toothed towards top and softly hairy beneath, mainly on veins.

FLOWERS AND FRUITS Flowers May in domed clusters, each flower about 1.5cm across. Fruit pear-shaped, green or brownish, 2–3cm long. Fruit edible only after being frosted.

HABITAT Widespread in southern Europe and Mediterranean region; endangered in some parts. Prefers dry deciduous woods. In Britain a rare native tree occurring naturally mainly in south Wales (Glamorgan), south-west England (Gloucestershire) and Bristol Channel area.

Wild Service
Sorbus torminalis

DESCRIPTION Height 15–25m. Deciduous. Young tree has a cone-shaped crown that becomes columnar and domed. May produce suckers. Bark dark brown and grey, smooth, fissures and plates in old trees. Leaves have wide basal lobes, bronze, yellow and russet in autumn, 6–14cm long.

FLOWERS AND FRUITS Flowers June with clusters of white flowers. Fruit globe-shaped, brown when ripe, in hanging clusters, 1–1.5cm long.

HABITAT Found in mixed woodland in southern and central Europe, including England north to Cumbria, to Asia Minor and North Africa. Fruits were once used to sweeten beer and to produce a potent alcoholic drink, 'checkers'; wood to make charcoal.

Common Hawthorn
Crataegus monogyna

DESCRIPTION Height 10–16m. Deciduous. Crown is rounded. Branches dense; twigs bear numerous sharp thorns. Leaves, 1.5–5cm long, have 2–3 pairs of lobes; shiny green turning yellow or red in autumn. Also called May and Quickthorn.

FLOWERS AND FRUITS Flowers May in clusters of creamy-white flowers. Fruit a small dark red berry, 8–14mm long, with one seed.

HABITAT Native to Europe, where it is widespread. Widely used as a hedging shrub, and also as a standard in woodlands. Berries a valuable food for birds such as Redwings and Fieldfares, and rodents such as Bank Voles; also foodplant of larvae of many moth species, notably those of the Lappet, and of the Hawthorn Shield Bug.

ASSOCIATED SPECIES

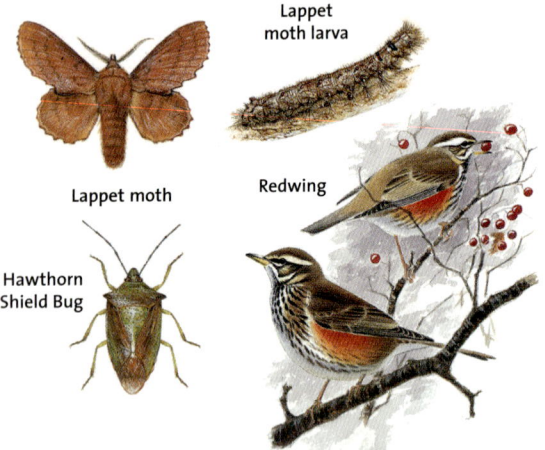

Lappet
moth larva

Lappet moth

Redwing

Hawthorn
Shield Bug

Crataegus calycina

DESCRIPTION Height to 11m.
Deciduous. Elongated
fruits most easily
distinguish this
woodland hawthorn
species from its relative,
Common Hawthorn
(page 120). Thorns few,
only 4–13mm long. Thin
broad leaves, 2.5–6.5cm
long; divided into
toothed lobes that
reach only a third of
the way to the midrib;
lowest lobe often
deeper than the others.

FLOWERS AND FRUITS
Flowers May–June.
Fruit dark or bright
red, roughly cylindrical,
containing a single
seed, 9–15mm long.

HABITAT Native to
deciduous and
evergreen woods in
northern and central
Europe, replacing
Common Hawthorn
in more extreme climates.

Midland Hawthorn
Crataegus laevigata

DESCRIPTION Height 12m. Deciduous. Rounded crown with dense branches. Bark grey- pink- or orange-brown, fissured and cracked. Leaves have 1–2 pairs of shallow lobes, uppersides shiny green, yellow in autumn, 1.5–6cm long.

FLOWERS AND FRUITS Flowers May in clusters of white or pink flowers. Bunches of red berries produced, 8 x 11mm long, containing 2–3 seeds.

HABITAT Native to central and western Europe, growing in heavy soil in woodland. Hybrids planted as street trees throughout Europe.

Blackthorn
Prunus spinosa

DESCRIPTION Height to 5m. Deciduous. Small tree or shrub. Rounded crown, and dense and wide-spreading branches. Suckers readily, forming scrub. Intricately branched twigs have sharp thorns and are downy when young. Bark black and rough. Leaves dull yellow-green, elliptic, 2–4.5cm long. Also called Sloe.

FLOWERS AND FRUITS Solitary white flowers appear March–April on previous year's growth, before leaves emerge. Fruit globe-shaped, 1.5cm long, dark blue to nearly black, with a fleshy outer covering concealing a large stone.

HABITAT Native throughout Europe except far north, growing in hedgerows and copses. Berries are eaten by birds, and also picked and used to flavour gin. Leaves are the foodplant for many moth larvae, and for the larvae of the scarce Brown and Black Hairstreak butterflies. Flowers are important for insects such as bumblebees in spring.

ASSOCIATED SPECIES

♀ Brown
Hairstreak

White-tailed
Bumblebee

Plum
Prunus domestica

DESCRIPTION Height 6–10m. Deciduous. Tree with rounded crown that suckers readily. Bark brown or grey-brown, becoming fissured. Leaves egg-shaped, 3–8cm long, wider towards tips.

FLOWERS AND FRUITS Flowers in early spring on previous year's growth, before leaves emerge. White flowers borne singly or in pairs. Fruit globe-shaped, 2–8cm long, dark blue-purple when ripe, a large stone surrounded by juicy sweet flesh. Edible and good.

HABITAT Native to the Caucasus, but extensively cultivated. Grows in orchards and gardens, and naturalized elsewhere.

Myrobalan Plum
Prunus cerasifera

DESCRIPTION Height 8–12m. Deciduous. Young tree has an upright crown that grows rounded and domed. Suckers grow from roots. Bark smooth, turning darker and with plates. Leaves oval, toothed, deep shiny green above, matt green below, 4–6cm long. Also called Cherry Plum.

FLOWERS AND FRUITS Flowers on previous year's growth in winter or early spring. Flowers are white. Edible fruit red or purple when ripe, 2.5–3cm long.

HABITAT Native to the Balkans to central Asia. Abundantly naturalized in Britain, growing in hedgerows and field margins.

Wild Cherry
Prunus avium

DESCRIPTION Height to 25m. Deciduous. Crown cone-shaped at first, then rounded. Bark grey-pink to red. Young trees have shiny bark that later becomes fluted with dark fissures. Leaves oval, 7–12cm long, margins serrated, dark green uppersides, paler below and downy in young leaves.

FLOWERS AND FRUITS Flowers April, in clusters with new leaves. Fruit globe-shaped, dark red or yellow-red, 2cm long, enclosing stone.

HABITAT Widespread throughout Europe, as well as North Africa and south-west Asia. Found in mixed woodland.

Bird Cherry
Prunus padus

DESCRIPTION Height 10–20m. Young trees have a cone-shaped crown, which tends to become domed. Branches droop. Bark smooth, grey-brown. Leaves elliptic or egg-shaped, 7–13cm long, wider towards tips, margins serrated. Dark green turning yellow and red in autumn.
FLOWERS AND FRUITS Flowers late spring on side shoots. Flowers white, on drooping racemes. Fruit black, globe-shaped, up to 8mm long.
HABITAT Widespread native across northern Europe to northern Asia. Absent from Mediterranean. Grows in temperate woodlands, preferring lime soils.

Sour Cherry
Prunus cerasus

DESCRIPTION Height to 8m. Deciduous. Round-headed tree or large shrub with a poorly defined short or branched trunk. Often produces suckers. Leaves slightly leathery, 3–8cm long, abruptly pointed at tips, small rounded teeth at margins. Paler prominently veined undersides downy at first, later smooth.

FLOWERS AND FRUITS Flowers April–May, just before leaves, in clusters of 2–6. Fruit 1.8cm long, bright red, long-stalked and acid-tasting.

HABITAT Native to south-west Asia. Introduced to Europe for its edible fruits, which are used mainly in preserves. Extensively planted and naturalized in Britain, generally by suckering.

Cherry Laurel
Prunus laurocerasus

DESCRIPTION Height to 15m. Evergreen. Crown domed and rounded, often spreading, with many stems. Bark black or grey-brown. Leaves large, 8–20cm long, lance-shaped, in sprays.

FLOWERS AND FRUITS Thirty to forty cream-coloured flowers are produced on erect racemes, in late winter on previous year's growth. Globe-shaped fruit ripens in autumn to dark purple.

HABITAT Native to the Balkan Peninsula. Widely grown as an ornamental shrub in much of Europe, and commonly naturalized in open woods.

Apricot
Prunus armeniaca

DESCRIPTION Height 3–10m. Deciduous. Small rounded tree with smooth reddish twigs and young leaves, and twisted branches. Leaves finely toothed, 5–10cm long, nearly circular, with abruptly pointed tips and straight or heart-shaped bases. Mature leaves dull green above, greenish-yellow beneath.

FLOWERS AND FRUITS Flowers March–April before leaves appear. Flowers short-staked, solitary or paired; hairy flower tube bell-shaped, white or pale pink, 1–1.5cm long. Fruit round, yellow to orange and downy, 4–8cm long.

HABITAT Native to central Asia and China. Widely grown and naturalized in southern Europe; rare outdoors in Britain.

Peach
Prunus persica

DESCRIPTION Height to 6m. Deciduous. In autumn large globular fruits with a red flush distinguish this bushy straight-branched tree from the similar Almond (page 134). Leaves lance-shaped, finely toothed, folded lengthways into V-shape, 5–15cm long. **FLOWERS AND FRUITS** Flowers March–May, usually solitary, appearing as leaf buds open. Flower tubes as broad as long, deep pink, sometimes pale pink or white, 2cm long. Fruit globular, yellow flushed with red, and downy, sweet and juicy, 4–8cm long. **HABITAT** Native to northern China. Long cultivated in southern Europe in orchards and gardens; occasionally grown in Britain.

Almond
Prunus dulcis

DESCRIPTION Height 10m. Deciduous. Broad crown. Smooth dark brown bark, growing fissured. Leaves lance-shaped with pointed tips, dark to yellow-green, 4–12cm long.

FLOWERS AND FRUITS Flowers late winter–early spring on previous year's growth. Large and showy bright pink or pinkish-white flowers, 2.5–5cm long. Fruit egg-shaped with green and fleshy, rather felt-like skin that splits to reveal the nut, 3.5–6cm long.

HABITAT Native to south-west Europe and North Africa, probably originating in Asia. Prefers dry hillsides. In warm climates widely cultivated for nuts; grown as an ornamental in north.

Tree of Heaven
Ailanthus altissima

DESCRIPTION Height 15–30m.
Deciduous. Branches
radiating, giving a
rounded crown, often
with suckers. Bark
smooth or scaly, grey-
brown with pale stripes.
Leaves, 60cm long, have an
unpleasant smell; they are
are alternate, 13–23 red-
stalked opposite leaflets,
deep green on top, pale
and hairless beneath.

FLOWERS AND FRUITS Male
and female flowers
open May, usually
growing on separate
trees. Fruits in late
summer, producing
winged seeds,
3.5–4cm long.

HABITAT Native to
northern China,
preferring dry light soils.
Occurs in forests in the wild.
Very tolerant of air pollution.
May be planted in cities.
Cultivated elsewhere and
often naturalized.

Carob
Ceratonia siliqua

DESCRIPTION Height to 10m. Evergreen. Domed and thickly branched, bushy low tree. Leaves pinnate with 2–5 pairs of leaflets and no terminal leaflet, margins may be wavy, 3–5cm long, dark, shiny green above, pale beneath.

FLOWERS AND FRUITS Flowers August–October, tiny, lacking petals, and grouped in short green spikes. Male and female flowers may be borne on the same or separate trees. Pods violet-brown when ripe, 10–20cm long.

HABITAT Native to dry areas of Mediterranean, where it is also grown for fodder.

Common Laburnum
Laburnum anagyroides

DESCRIPTION Height to 12m. Deciduous. Branches ascending and arching. Bark smooth, dark green at first, then brown. Leaves alternate, each leaf consisting of three leaflets, 3–8cm long, grey-green on upper surfaces, silky white hairs beneath.

FLOWERS AND FRUITS Yellow flowers are pea-like and poisonous, produced May–June, hanging in long chains, 10–30cm long. Pods, 4–6cm long, containing black seeds, hang on tree in winter.

HABITAT Native to southern and central Europe, growing in scrub and woodland. Widely planted as an ornamental, and often naturalized. The heartwood, which is very hard, has been used as a substitute for ebony.

Golden Wreath
Acacia saligna

DESCRIPTION Height to 10m. Evergreen. Small tree or shrub with weeping twigs. Leaves pendulous, long, and straight or sickle-shaped, dull or shiny with a single vein, variable in size and shape, usually 10–20cm long. Also called Orange Wattle.

FLOWERS AND FRUITS Flowers March–May, grouped into spherical heads, 1–1.5cm long. Pods straight, 6–12cm long, narrow and flattened, pinched between each black seed.

HABITAT Native to Western Australia and widely planted in southern Europe. Used as a sand stabilizer. Not very hardy in Britain, although it can tolerate occasional temperatures down to -5 to -10°C.

Silver Wattle
Acacia dealbata

DESCRIPTION Height
to 30m. Evergreen.
Feathery and delicate
with silvery-hairy
foliage, especially when
young. Bark grey-green
and smooth. Leaves
pinnately divided, each
division pinnately
divided into leaflets
5mm long; more bluish
above than below.

FLOWERS AND FRUITS
Flowers January–
March, with numerous
spherical pale yellow
heads, 5–6mm long.
Pods brown, 4–10cm
long, flattened and
slightly pinched
between seeds.

HABITAT Native to south-
east Australia and Tasmania. Planted in southern Europe mainly for
timber and as an ornamental. Hardiest acacia, locally frequent in
mildest areas of southern England. The 'mimosa' of florists. Widely
naturalized, especially in understory of woods.

Judas Tree
Cereis siliquastrum

DESCRIPTION Height 10–15m. Deciduous. Branches are spreading. Bark grey-brown, smooth or folded, becoming cracked. Leaves, 8–12cm long, alternating with heart-shaped bases, in two rows, wavy untoothed margins.

FLOWERS AND FRUITS Flowers on short stalks, singly or in cluster, in May before or with leaves. Flat pods, 10cm long, ripen to purple in autumn and remain on tree until well into winter.

HABITAT Native to Mediterranean, and southern Europe to western Asia. Planted as an ornamental in other parts of Europe, often growing in British parks and gardens in warmer areas.

Sycamore
Acer pseudoplatanus

DESCRIPTION Height 15–35m. Deciduous. Massive domed crown in mature trees. Bark grey and smooth, becoming scaly and fissured. Leaves, to 25cm long, have five pointed lobes, and are dark green on top, pale underneath.

FLOWERS AND FRUITS Flowers April–June, in a dense panicle hanging like catkins, to 12cm long. Fruit hairless with wings at 50–60-degree angle, to 6cm long. Two single seeds dispersed by wing.

HABITAT Native to western Europe, east to the Caucasus and Crimea, growing in woodland and forests. Naturalized elsewhere. Resistant to salt winds and air pollution. May be invasive.

Field Maple
Acer campestris

DESCRIPTION Height 10–12m, occasionally to 28m. Deciduous. Forms a rounded dome with grey-brown scaly bark that darkens in old trees. Leaves, 7cm long, have five lobes, may be pink or purplish when new, turning yellow to reddish-brown in autumn.

FLOWERS AND FRUITS Yellowish flowers are not very obvious, appearing after leaves, in April–May. Fruit is flat on a pair of wings, 2.5–5cm long, which help the seeds disperse by wind.

HABITAT Native to central and southern Europe, including England and Wales, growing in deciduous broadleaved woodland and hedgerows. Favours sandy, chalk and clay soils. Fallen seeds are a key source of food for Wood Mice and Bank Voles; leaves sustain the larvae of a number of moths. Timber was once used for turnery and carving.

ASSOCIATED SPECIES

Wood Mouse

Bank Vole

Box Elder
Acer negundo

DESCRIPTION Height to 20m. Deciduous. Numerous twiggy sprouts grow from trunk and branches of this fast-growing but short-lived tree. Leaves opposite, 10–15cm long, pinnately divided into toothed oval leaflets.

FLOWERS AND FRUITS Flowers March, appearing before leaves on separate male and female trees. Female flowers greenish; male flowers with conspicuous red anthers; no petals. Dry fruit, paired, 2cm long. Slightly curved wings form acute angle.

HABITAT Native to eastern North America. In Britain variegated forms are commonly planted as street trees and ornamentals; also occasionally naturalized.

Norway Maple
Acer platanoides

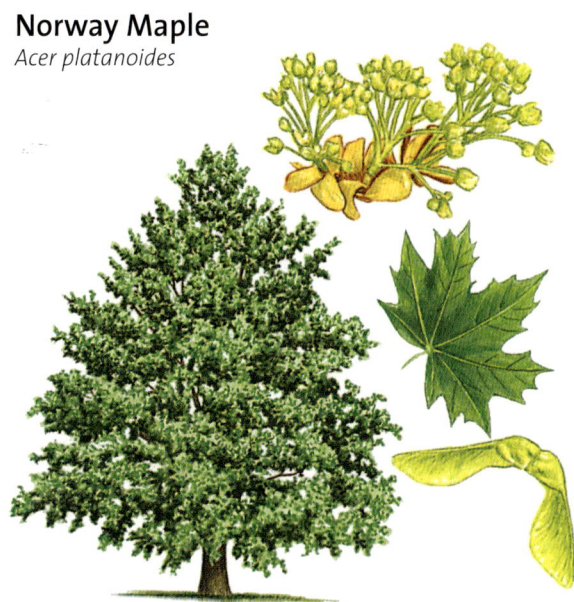

Description Height 15–25m. Deciduous. Young trees have an egg-shaped crown that becomes rounded or dome-shaped. Bark smooth, developing shallow fissures. Leaves, 7–14cm long, have five lobes; new leaves may be purplish, turning yellow in autumn.

Flowers and fruits Flowers usually yellow-green, appearing in early spring before leaves emerge. Fruit a flat nut with wings.

Habitat Native to upland forests and deciduous woodland in Europe, from southern Sweden and southern Norway to the Crimea. Widely planted, especially as an amenity tree in towns.

Horse Chestnut
Aesculus hippocastanum

DESCRIPTION Height 20–35m. Deciduous. Magnificent spreading tree with a domed crown, large divided leaves and showy flower candles. Bark red-brown or dark grey-brown, and scaly. Branches arching, usually turned up at ends. Large winter leaf buds, to 3.5cm long, extremely sticky. Leaves have 5–7 large stalkless leaflets, broader towards tips, each to 25cm long.

FLOWERS AND FRUITS Flowers are showy white panicles on spikes, to 30cm long, bearing 40 or more flowers that open in May. Globe-shaped spiny fruit contains one or more large shiny brown nuts (conkers), 3–5cm long.

HABITAT Native to the Balkan Peninsula. Planted as an ornamental elsewhere in Europe. Abundantly planted in Britain since the early 17th century. Grows in mixed woodland and parks. Flowers important to nectar-seeking insects, particularly bees.

ASSOCIATED SPECIES

Honey Bee

Holly
Ilex aquifolium

DESCRIPTION Height 10–25m. Evergreen. Striking narrow-crowned conical tree with bright red berries. Bark green when young, smooth and grey later. Leaves alternate with sharp spines, to 12cm long, dark green and glossy on top, pale green beneath, and hairless. Leaves tend to become smoother, less spiky, as tree ages, especially near top of crown, just retaining end point.

FLOWERS AND FRUITS Male and female flowers appear in May on separate trees. Pollinated by insects. Only female trees bear berries.

HABITAT Native to western Europe across to western Asia. Grows in scrub, woodland and hedgerows. Many cultivars – hollies are often used as ornamental garden trees and for hedging. In winter, berries are popular with thrushes such as the Mistle Thrush; and buds are a food of the first-generation larvae of the Holly Blue butterfly. Leaves were once used as winter feed for livestock, while wood was used in the making of whips.

ASSOCIATED SPECIES

Holly Blue

Mistle Thrush

Spindle
Euonymus europaeas

DESCRIPTION Height 6–10m. Deciduous. Normally forms a slender twiggy tree or shrub. Crown egg-shaped; branches ascending. Bark fissured, grey-brown in colour. Leaves elliptical, to 10cm long, tapering, medium to dark green in summer, turning pinkish-red in autumn.
FLOWERS AND FRUITS Three to five small yellow-green flowers borne May–June. Pendent fruit has four lobes, each with a single orange seed, and is pink-red, 1.2–1.5cm long. Fruit is poisonous.
HABITAT Native to most of Europe except far south and far north. Grows in woodland, hedgerows and scrub. Prefers chalk and lime soils. Often planted as an ornamental. Wood from the tree was once used to make spindles.

Box
Buxus sempervirens

DESCRIPTION Height 6m. Evergreen. Spreading shrub or narrow tree. Bark pale brown, cracking into squares. Leaves opposite in sprays, oval, mid to dark green, 1.2–2.5cm long.

FLOWERS AND FRUITS Clusters of 5–8 male flowers and one female flower at the end produced late spring. Fruit almost globe-shaped, blue-green, 7mm long. Seeds black and shiny.

HABITAT Native to central and southern Europe. In Britain widely planted as a hedging shrub, especially in formal gardens. Prefers brick soils on dry hills. Can withstand dense shade.

Alder Buckthorn
Frangula alnus

DESCRIPTION Height 6m. Deciduous. Shrub or small tree with rounded crown with several stems, and bark that is dark-brown, showing yellow beneath if cut. Leaves oval, 3–7cm long, tapered, with wavy margins.

FLOWERS AND FRUITS Flowers May–June; small clusters of flowers formed in leaf axils. Fruit globe-shaped and berry-like, ripening from green through yellow to purplish-black, 6–10mm long.

HABITAT Native to central and northern Europe, including Britain; absent from far north and parts of Mediterranean. Grows in hedgerows and understorey in moist woodland. Leaves are particularly attractive to the larvae of Brimstone butterflies, while flowers attract bees and birds feed on berries. Wood was once used to make butchers' spikes and skewers.

ASSOCIATED SPECIES

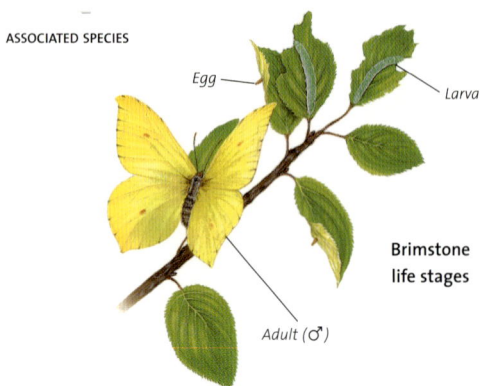

Egg

Larva

Brimstone life stages

Adult (♂)

Purging Buckthorn
Rhamnus catharticus

DESCRIPTION Height 6m.
Deciduous. Small tree to
large shrub with rounded
crown and brown bark.
Leaves elliptical, dull
green upper surfaces,
pale green beneath,
2.5–7cm long.

FLOWERS AND FRUITS
Flowers May–June,
unobtrusive yellow-
green, fragrant,
4mm across.
Male and female
flowers grow on
separate trees. Fruit
almost globe-shaped,
purple-black, 6mm
long. It gives the tree
its purgative qualities.

HABITAT Native to Europe
and northern Asia.
Found on sandy scrub
and chalk downland.

Common Lime

Tilia x europaea

DESCRIPTION Height 20–45m. Deciduous. Crown a tall dome. Bark smooth and dull grey, becoming brown and ridged, with much side growth. Leaves broader at bases, to almost round, 6–15cm long, toothed, often shiny and sticky with honeydew. **FLOWERS AND FRUITS** Highly scented hanging flowers in July, 4–10 per bract; yellowish-white petals. Fruit egg- or globe-shaped, hairy, with a pointed tip, hangs in clusters. **HABITAT** Hybrid between Small- and Broad-leaved Limes (pages 156–8). Widely planted and very common in parks and towns.

Small-leaved Lime
Tilia cordata

Description Height 20–25m. Deciduous. Crown a tall and rather narrow dome. Branches ascend, but their tips point downwards. Bark smooth and grey in young trees, while old trees have scaly bark near base and become dark grey or brown. Leaves rounded to egg-shaped, toothed, blue-green, to 9cm long.

Flowers and fruits Flowers mid-summer, fragrant, whitish-cream petals, in clusters of up to 10 on green bract 10cm long. Fruit globe-shaped or elliptical, about 2mm diameter.

Habitat Native of central Europe, including Britain, Norway and Sweden, to the Caucasus. Grows in woodland. Widely planted in towns and also used for forestry. Larvae of Lime Hawkmoths feed mainly on the leaves of lime trees (the adults do not feed). Large populations of aphids on limes result in the drip of honeydew from the trees.

ASSOCIATED SPECIES

Lime Hawkmoths mating

Broad-leaved Lime
Tilia platyphyllos

Description Height 20–40m. Deciduous. Young tree has ascending branches and a narrow crown. Leaves heart-shaped, hairy and toothed, 6–15cm long. Also called Large-leaved Lime.

Flowers and fruits Flowers mid-summer with whitish-yellow, five-petalled, fragrant and pendent flowers, which are attached to a wing-like bract. Fruit hard and woody, usually spherical, pendent, to 1.8cm long.

Habitat Native to Europe (including England) through to Asia Minor. Favours lime-rich soils. Grows in woodland. Widely cultivated. Also grown as an ornamental tree.

Tamarisk
Tamarix gallica

DESCRIPTION Height to 8m.
Deciduous. Crown rounded.
Bark brown with fissures and
scales. Leaves fine, to 6mm long,
narrow and frond-like, giving a
delicate feathery appearance to
this small tree.

FLOWERS AND FRUITS Flowers in
summer on short stalks
carrying 5–10cm racemes
of pink-white flowers.
Fruit a small capsule
containing many
seeds with hairs at
one end.

HABITAT Native to
coastal north-west
France across to North
Africa. Resistant to salt
winds and in Britain often
planted as an ornamental,
usually near the sea. Also
widespread as a small garden
tree and shrub.

Oleaster
Eleagnus angustifolia

DESCRIPTION Height 6–12m. Deciduous. Grows as a shrub or small tree. Rounded crown and fissured bark. Leaves lance-shaped, underside silvered, 2.5–9cm long. Also called Russian Olive.

FLOWERS AND FRUITS Flowers June, bearing fragrant bell-shaped flowers that occur singly or sometimes in groups of 2–3. Oval-shaped fruit is edible and ripens to yellow, 1–1.5cm long.

HABITAT Native to western Asia. Naturalized in southern and central Europe. Grows on open sites. Sometimes planted in British gardens.

Sea Buckthorn
Hippophae rhamnoides

DESCRIPTION Height 13m. Deciduous. Grows as a shrub or small tree. Suckers grow from roots. Bark ridged, brown in colour. Leaves slender, silver-grey, 2.5–7.5cm long.

FLOWERS AND FRUITS Flowers March–April on separate male and female trees. Fruits on female plants are orange berries, 6–8mm long, each holding a single seed.

HABITAT Native to western Europe, including coastal eastern England, across to Asia and western China. Grows in coastal dunes and sandy areas. Planted as a sand binder and ornamental, and often naturalized on open sites.

Silver Gum
Eucalyptus cordata

DESCRIPTION Height 35m.
Evergreen. Elegant plant
grown as a tree or pruned
shrub. It can also be coppiced.
Bark smooth, greenish-white
and blotched. Leaves heart-shaped,
bluish-white, glaucous.
FLOWERS AND FRUITS Flowers fragrant,
large and white. Fruit large.
HABITAT Native to south-east
Tasmania. Grows best in fertile
acid to neutral soils. Grown for
timber. Elsewhere grown as an
ornamental. A comparatively
tender tree that will not withstand
hard frost. Blue foliage is valued
by florists.

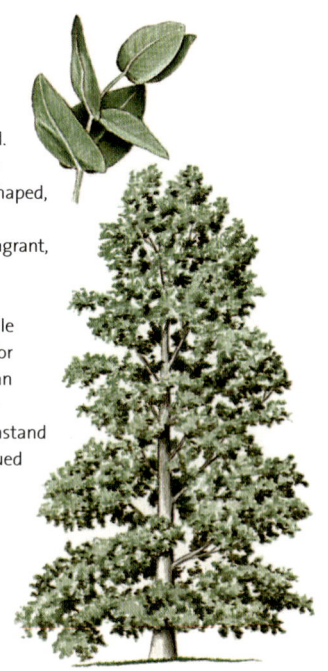

Cider Gum
Eucalyptus gunnii

DESCRIPTION Height 20–30m.
Evergreen. Domed crown in
mature trees. Bark peels in big
flakes showing paler patches,
which change to grey. Leaves
in two shapes: young leaves
rounded or oval, green or blue-
green, opposite, to 4cm long;
adult leaves alternate, long and
green, to 7cm long.

FLOWERS AND FRUITS Flowers
January–February in groups of
three. Fruit egg-shaped, to 1cm long;
ripens in summer with new flowers.

HABITAT Native to Tasmania and south-
east Australia, growing in mountainous
forest. Fast-growing species widely planted
in cooler parts of Europe, mainly for its
ornamental juvenile foliage. The most
planted species of eucalyptus in Britain.

Blue Gum
Eucalyptus globulus

DESCRIPTION Height 10–45m. Evergreen. Crown cone-shaped at first, maturing into a tall dome. Bark peels to reveal white, which eventually becomes grey-brown. Juvenile leaves lance-shaped to oval, glaucous, 7–16cm long; adult leaves lance-shaped and curved, leathery, blue-green, 10–30cm long.
FLOWERS AND FRUITS Single flowers are formed. Fruit is globe-shaped and warty.
HABITAT Mainly native to Tasmania, but also to Victoria, Australia. Grown as an ornamental tree and for timber. Scarcely hardy in England, but common in Ireland and further south. A fast-growing tree often used for timber.

Mountain Gum
Eucalyptus dalrympleana

DESCRIPTION Height 20–35m. Evergreen. Fast-growing with a cone-shaped crown. Bark peels to reveal cream patches. Adult leaves lance-shaped and curved, blue-green, 10–18cm long. Juvenile leaves oval or rounded. Also called Broad-leaved Kindling Gum.

FLOWERS AND FRUITS Umbels of three flowers. Fruit is shaped like a hemisphere.

HABITAT Native to Tasmania, and also Victoria and New South Wales in Australia. Widely grown as an ornamental throughout Europe.

Red Gum

Eucalyptus camuldulensis

DESCRIPTION Height 20–80m. Evergreen. Broad-spreading crown with pendulous branchlets. Bark flaky, showing patches of yellow or grey. Two kinds of leaves: juvenile leaves lance-shaped, bluish, 6–9cm long; adult leaves narrow, lance-shaped, curved, glaucous to pale green, 12–22cm long.

FLOWERS AND FRUITS Flowers December–February; umbels consist of clusters of 5–10 white flowers. Fruit is like a hemisphere, 5–6cm long.

HABITAT Native to Australia. Widely grown for timber in temperate areas across Europe; also grown as an ornamental.

Pomegranate
Punica granatus

DESCRIPTION Height to 8m. Deciduous. Crown rounded. Bark brown, peeling off to show whitish or buff beneath. Narrow oval leaves, 3.5–7cm long.

FLOWERS AND FRUITS Bright red flowers occur singly or in small groups on current season's growth from June onwards. Fruit globe-shaped with a hard rind, and calyx remains attached. Each fruit contains many semi-transparent pink seeds. Edible and good.

HABITAT Native to south-west Asia. Widely cultivated in Mediterranean and other warm temperate areas. In Britain grows only in sheltered south-facing gardens, and in conservatories or greenhouses.

Pacific Dogwood
Cornus nuttallii

DESCRIPTION Height 15m. Deciduous. Conical in shape. Bark purple-brown and square-cracking. Leaves oval, mid to dark green, 12cm long. Brilliant red autumn colours to leaves.

FLOWERS AND FRUITS Flowers late spring. Four to six, sometimes eight, large creamy bracts, which develop a pink flush, in late spring, with sometimes more in autumn. Fruits spherical and orange-red.

HABITAT Native to western North America. Prefers well-drained neutral to acid soils. In Europe grows in large gardens in warmer areas.

Common Dogwood
Cornus sanguinea

Description Height 6m. Deciduous. Crown rounded. Bark brown. Oval leaves dark green, 4–8cm long. Autumn colour red or purple. The blood-red shoots give this species its scientific name, *'sanguinea'*.

Flowers and fruits Fragrant dull white flowers appear in June. Fruit globe-shaped, turning from green to almost black, 6–8mm long, with a single seed.

Habitat Native to southern Scandinavia, Britain and Portugal, across Europe to Turkey. Grows in understorey in woodland, hedgerows and woodland edges.

Strawberry Tree
Arbutus unedo

DESCRIPTION Height 8–12m. Evergreen. Trunk short, often leaning, with domed crown. Branches often ascending and twisted. Bark dark red, shredding as tree ages. Leaves elliptic or egg-shaped, wider near tips, dark green, leathery, toothed, 5–10cm long.

FLOWERS AND FRUITS Flowers October–December with drooping clusters of 15–20 white or pinkish flowers 9mm long. Edible but not very tasty globe-shaped fruits, 1.5–2cm across. They mature over 12 months, from green and yellow to bright red, ripening in autumn as the flowers are opening.

HABITAT Patchy distribution from south-west Ireland, southern Europe, to Cyprus and Turkey. Tolerates damp acid soils to chalky hillsides. In warmer parts of Britain planted in large gardens and towns as an ornamental. Birds are fond of the fruits. In the Mediterranean the larvae of the spectacular Two-tailed Pasha – the largest European butterfly species – are entirely dependent on the leaves of the tree as a foodplant.

ASSOCIATED SPECIES

Two-tailed
Pasha

Rhododendron
Rhododendron ponticum

DESCRIPTION Height 10m. Evergreen. Usually on several stems to 15cm diameter, combining to give spreading crown. Bark brown. Leaves lance-shaped, shiny dark green, 10–20cm long.

FLOWERS AND FRUITS Flowers late spring–early summer, with large blossoms of light purple to pink-purple. Fruit a woody capsule, 2.5cm long.

HABITAT Native to Asia and possibly parts of Europe, where it is widespread. Naturalized in some areas of Britain. Difficult to eradicate and may smother other plants. Grows on hillsides and in shaded woodland.

Olive
Olea europaea

DESCRIPTION Height to 15m. Evergreen. Trunk short with a rounded spreading crown and silver-grey bark. Leaves lance-shaped, grey-green, whitish below, 2–8cm long.

FLOWERS AND FRUITS Flowers July–August on current season's growth. Flowers fragrant and produced in panicles. Egg- or globe-shaped fruits take a year or so to ripen from green to brown or black, 1–3.5cm long. Edible and good.

HABITAT Probably originated in south-west Asia and Saudi Arabia, where it was a key crop in ancient times, but has long been grown on hillsides across the Mediterranean region. Grown in gardens in warm areas of southern Britain – needs long and hot summers and mild winters to produce fruits.

Narrow-leaved Ash
Fraxinus angustifolia

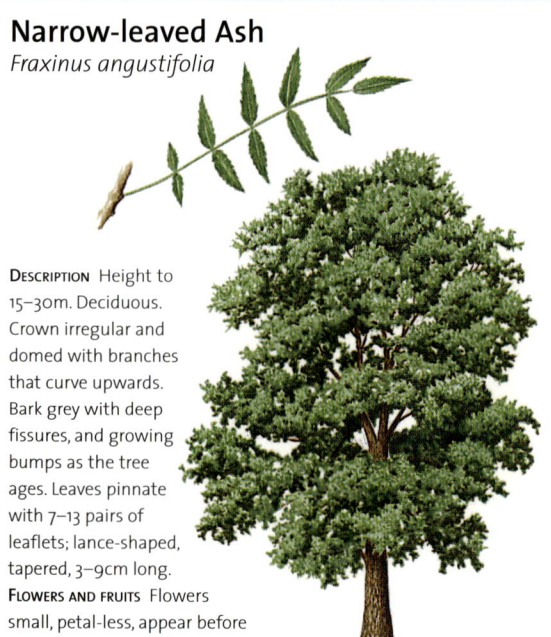

DESCRIPTION Height to 15–30m. Deciduous. Crown irregular and domed with branches that curve upwards. Bark grey with deep fissures, and growing bumps as the tree ages. Leaves pinnate with 7–13 pairs of leaflets; lance-shaped, tapered, 3–9cm long.
FLOWERS AND FRUITS Flowers small, petal-less, appear before leaves. Fruit a winged nutlet, 2–4.5cm long; hangs in small clusters, green maturing to brown.
HABITAT Native to western Mediterranean and North Africa, growing near rivers and in deciduous woodland. In Britain planted as an ornamental.

Manna Ash
Fraxinus ornus

Description Height 15–25m. Deciduous. Usually has a rounded crown with spreading branches. Bark dark grey, smooth. Leaves pinnate, 20–30cm long, comprising 5–9 leaflets that are variable in shape.
Flowers and fruits Flowers differ markedly from those of Common Ash (page 176). White to creamy fragrant flowers, about 6mm long, appear in May, in panicles about 20cm across. Fruit a winged nutlet, to 2cm long, hanging in dense clusters.
Habitat Native to southern Europe through to Asia Minor. Long grown in Britain as an ornamental, and very frequent in streets and parks.

Common Ash
Fraxinus excelsior

DESCRIPTION Height to 40m. Deciduous. Upper branches ascending, while lower ones curve downwards sometimes almost vertically, turning upwards near tips. Bark green-grey, fissuring with age. Leaves pinnate, 20–35cm long, with 9–13 toothed leaflets.

FLOWERS AND FRUITS Male and female flowers occur on the same tree, on separate twigs, in purplish clusters during April–May. Fruits are long seed pods that hang down in dense clusters. Winged seeds are called 'keys'.

HABITAT Native to Europe east to the Caucasus, including Britain and Ireland, growing in woodland, forests and hedgerows. Light and airy canopy allows a rich ground flora, comprising plants such as Bluebells and Wood Anemones, to flourish at tree bases. Valuable timber tree, the white wood of which was once used to make carriage shafts, and popular in the making of furniture and farm implements.

ASSOCIATED SPECIES

Bluebell

Wood
Anemone

Glossy Privet
Ligustrum lucidum

DESCRIPTION Height to 15m. Evergreen. Grown as a dense garden hedge, but forms a tree when left unclipped. Spreading branches; twigs flecked with white pores. Leaves opposite, thick, dark and very glossy above, matt beneath, 8–12cm long.

FLOWERS AND FRUITS Flowers August–September or later, in loose cones, 12–20cm long, small and white, with a heavy scent. Oval berries black with a white bloom, about 1cm long.

HABITAT Native to China. Widely used as a hedge, ornamental and street tree.

Wild Privet
Ligustrum vulgare

DESCRIPTION Height to 5m.
Evergreen, but some leaves fall in
cold weather. Branches long, arching
over and rooting on contact
with soil, making thickets.
Leaves lance-shaped,
opposite, 2–10cm long,
on short stalks.

FLOWERS AND FRUITS
Flowers June–July,
4–5mm across, white in
terminal panicles. Fruit
a poisonous, shiny
purple-black berry, 7–8mm
long, containing 2–4 seeds.

HABITAT Widespread across Europe.
Prefers lime and chalk soils. In Britain
locally common as a native species,
but generally scarce. Also widely
planted and naturalized, which obscures its natural range. Often used as
a hedging shrub. Berries are eaten by birds such as thrushes; leaves are
the foodplant of the striking larvae of Privet Hawkmoths.

ASSOCIATED SPECIES

Privet
Hawkmoth

Privet
Hawkmoth
larva

Lilac
Syringia vulgaris

DESCRIPTION Height 3–7m. Deciduous. Tree or shrub that suckers freely and often forms tall and very twiggy thickets. Leaves in opposite pairs, may be oval or heart-shaped; yellowish and slightly leathery, 4–8cm long.

FLOWERS AND FRUITS Flowers May–June, with shoots tipped with showy conical flowerheads that are intensely fragrant, 10–20cm long. Flowers fleshy, tubular, usually lilac, but may be white or cream in garden plants. Fruit capsule oval and brown, about 1cm long.

HABITAT Native to rocky scrub-covered hillsides in the Balkan Peninsula, but widely cultivated as an ornamental and for its attractive highly scented flowers. Naturalized in parts of Europe outside the Balkans, including in Britain.

Indian Bean
Catalpa bignonioides

Description Height 10–15m. Deciduous. Bark smooth, pink and brown in young trees, becoming grey and scaly with plates. Leaves egg-shaped, 12–25cm long, with slender pointed tips.

Flowers and fruits Flowers showy with tube-shaped white petals that have red and purple inner markings, in panicles 20–30cm long. Fruit, containing many winged seeds, is a purple pod, 15–40cm long.

Habitat Native to southern states of North America. Grows along woodland edges. Widely grown as an ornamental, including in many large cities in Britain.

Wayfaring Tree
Viburnum lantana

DESCRIPTION Height 6m. Deciduous. Small spreading tree or shrub with a rounded bushy crown. Bark brown. Leaves heart-shaped at bases, opposite and sharply toothed with white hairs beneath, 5–12cm long.
FLOWERS AND FRUITS Flowers May–June. Flowers all alike and fertile, forming a dense cream-coloured dome. Fruit an oval red berry, produced July–September, and turning to black with maturity, 8mm long.
HABITAT Native to southern and central Europe to Britain and Sweden. Grows in hedgerows and fringes of woods. In Britain common and widespread on waste gound and in soil with high nitrogen content.

Elder
Sambucus nigra

Description Height 10m. Deciduous. Crown rounded. Bark light brown or grey, thick, deeply fissured and corky. Many branches arise from base, later arching back towards the ground. Leaves opposite with terminal leaflet, 5–7 pairs of lateral leaflets, each to 12cm long.

Flowers and fruits Flowers creamy-white, 5mm across, massed in large clusters with flat-topped heads to 24cm across, and with a heavy sweet scent. Purple berries ripen August–September.

Habitat Common in Europe except far north, in hedgerows and woods. Flowers popular with nectar-feeding insects, berries with birds.

Laurustinus
Viburnum tinus

DESCRIPTION Height to 7m.
Evergreen. Flowers in
winter, bearing attractive
pink and white flowers.
Leaves opposite, entire
and narrowly to very
broadly oval in outline,
dark, slightly glossy and
thinly hairy beneath,
3–10cm long.
FLOWERS AND FRUITS Flowers in
February, and may continue
to June, in clusters, each
flower pale pink outside,
white within, 4–9cm long.
Fruit small and metallic
blue in colour.
HABITAT Native to
Mediterranean, but
often grown as an
ornamental elsewhere,
including in Britain,
where it is occasionally
naturalized. In natural
habitat grows in woods
and thickets on stony soils.

Guelder Rose
Viburnum opulus

DESCRIPTION Height 4m. Deciduous. Small spreading tree with few branches. Leaves opposite with 3–5 lobes, smooth and hairless on upper surfaces, hairy beneath, 5–10cm long. Leaves turn deep wine-red in autumn.

FLOWERS AND FRUITS Flowers June–July, bearing dense heads of small white fertile flowers, surrounded by bigger showy sterile flowers, 1.5–2cm across. Fruits round, ripening to translucent red September–October, hanging in clusters.

HABITAT Native across much of Europe, growing in damp woodland, hedgerows and thickets. Widespread in Britain.

Cabbage Palm
Cordyline australis

DESCRIPTION Height to 13m. Evergreen. Palm-like with bare forked trunks topped with dense tufts of narrow sword-shaped leaves. Trunks fork after flowering and often sucker, forming clumps. Leaves hard and sharp-pointed, dark green often tinged yellow, 30–90cm long.

FLOWERS AND FRUITS Flowers June–July, about 1cm across, creamy-white and fragrant, forming huge branched clusters on large spikes, to 1.2m long, growing from centre of crown. Berries blue-white, 6mm across.

HABITAT Native to New Zealand. Quite hardy and a popular ornamental and street tree in coastal areas of southern and western Europe, including Britain.

European Fan Palm
Chamaerops humilis

Description Height to 2m. Dwarf palm with stiff fan-shaped leaves, often on a thick fibre-covered trunk that rarely exceeds 2m. Leaves 1m long, green, greyish or bluish, deeply divided into narrow segments. Old leaf bases persist among white or grey fibres in trunk.

Flowers and fruits Flowers March–June, yellow, in a dense spike hidden among leaves; male and female flowers usually on separate trees. Globose fruits, either yellow or brown, 4.5cm long.

Habitat The only common native palm in Europe, mainly found in sandy coastal regions of western Mediterranean.

Chinese Windmill Palm

Trachycarpus fortunei

DESCRIPTION Height to 14m. Brown and shaggy trunk covered with matted fibrous bases of dead leaves, particularly on upper part near crown. Leaves fan-shaped, to 1m long, divided almost to bases into stiff and narrow pleated segments. Long leaf stalks are fibrous at bases.

FLOWERS AND FRUITS Fragrant egg-yolk yellow-coloured flowers March–June in many-branched conical clusters. Fruit three-lobed, purplish-white, about 2cm long.

HABITAT Native to China. Used as an ornamental along roads and avenues, mainly in Mediterranean region, but quite hardy and grown as far north as southern England.

Canary Island Date Palm
Phoenix canariensis

DESCRIPTION Height to 20m. Trunk, 1.5m in diameter, stout, unbranched and scarred from old leaf-stalks; bears a dense crown of many feathery fronds. Leaves to 6m long, pinnately divided into numerous narrow leaflets and viciously spiny towards bases of stalks.

FLOWERS AND FRUITS Flowers March–May; massive creamy-yellow clusters reach 2m long. Equally large clusters of dry and inedible orange fruits hang down from crown.

HABITAT Native to the Canary Islands. Often planted as an ornamental in streets, parks and leisure areas in Mediterranean region and south-west Europe. In Britain now frequent as a young plant in warm places.

Index